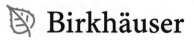

Progress in Mathematical Physics

Volume 72

More information about this series at http://www.springer.com/series/4813

Costas Bachas • Bertrand Duplantier • Vincent Rivasseau
Editors

The H Boson

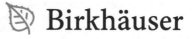 Birkhäuser

Editors
Costas Bachas
École Normale Superieure
Laboratoire de Physique Theorique
Paris, France

Bertrand Duplantier
Institut de Physique Théorique
CEA/Saclay
Gif-sur-Yvette Cedex, France

Vincent Rivasseau
Université Paris-Sud
Laboratoire de Physique Théorique
Orsay, France

ISSN 1544-9998   ISSN 2197-1846   (electronic)
Progress in Mathematical Physics
ISBN 978-3-319-86156-2   ISBN 978-3-319-57409-7   (eBook)
DOI 10.1007/978-3-319-57409-7

Mathematics Subject Classification (2010): 81-02

Printed on acid-free paper

This book is published under the trade name Birkhäuser, www.birkhauser-science.com
The registered company is Springer International Publishing AG
The registered company address is: Gewerbestrasse 11, 6330 Cham, Switzerland

Contents

Louis Fayard
Future Searches on Scalar Boson(s)

Abdelhak Djouadi
Implications of the H Boson Discovery

Contributors

François Englert
Service de Physique Théorique, Université Libre de Bruxelles, Bruxelles, Belgium

Yves Sirois
LLR, École Polytechnique, CNRS-IN2P3

Pierre Fayet
Laboratoire de Physique Théorique de l'École Normale Supérieure, Paris, France

Louis Fayard
LAL, Université Paris-Sud, Orsay, France

Abdelhak Djouadi
LPT, CNRS & Université Paris-Sud, Orsay, France

Séminaire Poincaré XIX

Le Boson H

Samedi
29 novembre 2014

F. Englert : Le boson de Brout - Englert - Higgs • *10h*

Y. Sirois : La découverte du boson H au LHC • *11h*

P. Fayet : Bosons scalaires et supersymétrie • *14h*

L. Fayard : Recherches futures sur les bosons scalaires • *15h*

A. Djouadi : Implications de la découverte du boson H • *16h*

INSTITUT HENRI POINCARÉ • Amphi Hermite
11, rue Pierre et Marie Curie • 75005 Paris

www.bourbaphy.fr

ÉCOLE POLYTECHNIQUE
UNIVERSITÉ PARIS-SACLAY

cea

TRIANGLE DE LA PHYSIQUE

FONDATION IAGOLNITZER

Foreword

This book is the sixteenth in a series of Proceedings for the *Séminaire Poincaré*, which is directed toward a broad audience of physicists, mathematicians, and philosophers of science.

The goal of the Poincaré Seminar is to provide up-to-date information about general topics of great interest in physics. Both the theoretical and experimental aspects of the topic are covered, generally with some historical background. Inspired by the *Nicolas Bourbaki Seminar* in mathematics, hence nicknamed *"Bourbaphy"*, the Poincaré Seminar is held on a yearly basis at the Institut Henri Poincaré in Paris, with written contributions prepared in advance. Particular care is devoted to the pedagogical nature of the presentations, so that they may be accessible to a large audience of scientists.

This new volume of the Poincaré Seminar Series, **The H Boson**, in the Birkhäuser Series *Progress in Mathematical Physics* corresponds to the nineteenth such seminar, held on November 29, 2014, at Institut Henri Poincaré in Paris. Its first aim is to provide a detailed description of the seminal theoretical construction in 1964, independently by Robert Brout and François Englert, and by Peter W. Higgs, of a mechanism for short-range fundamental interactions, now called the *Brout–Englert–Higgs (BEH) mechanism*. It accounts for the non-zero mass of elementary particles and predicts the existence of a new particle – an elementary massive scalar boson. The second aim of the present volume is then to describe the experimental discovery of this fundamental missing element in the *Standard Model* of particle physics. The H boson, also called the Higgs boson, was produced and detected in the Large Hadron Collider (LHC) of CERN near Geneva by two large experimental collaborations, ATLAS and CMS, which announced its discovery on the 4th of July 2012.

The Nobel Prize in Physics 2013 was awarded jointly to François Englert and Peter W. Higgs *"for the theoretical discovery of a mechanism that contributes to our understanding of the origin of mass of subatomic particles, and which recently was confirmed through the discovery of the predicted fundamental particle, by the ATLAS and CMS experiments at CERN's Large Hadron Collider"*. This does not seem to preclude a later awarding of another Nobel Prize in Physics, this time to the CERN organization, for such an experimental achievement of historical importance. The present volume is a testimony to the extraordinary fifty-year long scientific adventure which led to the discovery of the H boson.

Englert's and Higgs' Nobel Lectures, delivered on 8 December 2013 at Aula Magna in Stockholm University[1] were introduced by Lars Brink, Chairman of the Nobel Committee for Physics, who also gave the Presentation Speech at the Award Ceremony on 10 December 2013. Brink was in charge for the Class for Physics of the Royal Swedish Academy of Sciences of the very detailed report on the Scientific Background on the Nobel Prize in Physics 2013, "The BEH-Mechanism, interactions with short-range forces and scalar particles", which can be found on the internet[2].

The first article in this volume, entitled "The BEH Mechanism and its Scalar Boson", is the text of the Nobel Lecture given by François ENGLERT. Using the example of the effective thermodynamical potential of a ferromagnet above and below the Curie point, the author explains how, at a given minimum, the curvature of the effective potential measures the inverse susceptibility, which is the analog of mass in relativistic particle physics. At non-zero magnetization, the curvature vanishes along rotations of the order parameter, while it is positive in the longitudinal direction. As discovered by Y. Nambu on 1960 in the context of the Bardeen-Cooper-Schrieffer (BCS) theory of superconductivity, and established generally by J. Goldstone, such massless modes are the characteristic signature of *spontaneously broken continuous symmetries*. Inspired by BCS theory, where the quantum phase symmetry is broken by a condensation of electron pairs bound by an attractive force due to phonon exchange, Nambu introduced spontaneous symmetry breaking (SSB) in relativistic quantum field theory, and showed how a fermion condensate could break the (approximate) chiral symmetry of strong interactions, leading to the emergence of (nearly) massless pseudoscalar particles called *pions*. He was awarded half of the 2008 Nobel Prize in Physics *"for the discovery of the mechanism of spontaneous broken symmetry in subatomic physics"*, while the other half was awarded jointly to M. Kobayashi and T. Maskawa *"for the discovery of the origin of the broken symmetry which predicts the existence of at least three families of quarks in nature"*. However, as shown by P.W. Anderson in 1963, in superconductivity the presence of long-range Coulomb interactions converts the massless Nambu–Goldstone (NG) mode into a "massive" plasmon mode of electron density oscillations, as the longitudinal partner of transverse electromagnetic modes. F. Englert explains why NG bosons similarly no longer exist when the spontaneously-broken symmetry is gauged to a *local* symmetry, a fact understood in 1964 studies by Brout and Englert, by Higgs, by Migdal and Polyakov (delayed in USSR to 1965), and by Guralnik, Hagen and Kibble. Via their coupling to the gauge fields, the NG massless bosons are replaced by extra polarizations for the spin-1 gauge bosons which, by relativistic invariance, must be massive. This BEH mechanism accommodates in the same theory both long-range and short-range interactions, by leaving part of the gauge symmetry unbroken and the

---

[1] see http://www.nobelprize.org/nobel_prizes/physics/laureates/2013/

[2] see http://www.nobelprize.org/nobel_prizes/physics/laureates/
       2013/advanced-physicsprize2013.pdf

corresponding vector bosons massless. The most impressive success of these ideas is the unification of weak and electromagnetic interactions within the *Standard Model* of particle physics. Its validity required the discovery of one missing, new massive particle called the *H boson*, the analog of positive-curvature fluctuations in the ferromagnetic picture.

The second article in this volume, "Discovery and Measurements of the H Boson with ATLAS and CMS Experiments at the LHC", is written by Yves SIROIS, the leader for France of the CMS collaboration. The author recollects the adventure of the search for the Higgs boson, starting with the LEP e^+e^- collider in the nineties, continuing at the Tevatron $p\bar{p}$ collider and culminating with the recent discovery at LHC. This long road underlines one of the important challenges of the search, the absence of a theoretical prediction for the H boson's mass. Sirois then reviews the measurements of various properties of the new particle (spin, decay modes, etc.), all of which are, up to now, compatible with those expected for the Higgs boson of the Standard Model.

The third contribution, entitled "Scalar Bosons and Supersymmetry", is written by Pierre FAYET, a theoretical physicist at École Normale Supérieure in Paris, who is one of the pioneers of this tanatalizing new symmetry, which relates fermions and bosons and arises in unified theories such as Superstring Theory. The author places the H boson discovery in the larger context of possible extensions of the Standard Model involving supersymmetry. An extended Higgs sector, involving four or more extra scalar particles is predicted by supersymmetric theories, as well as other experimental signals for which the LHC collaborations are actively searching.

In "Future Searches on Scalar Boson(s)", Louis FAYARD, a prominent member of the ATLAS Collaboration, reviews the prospects for other experimental discoveries, both at LHC and in future colliders. Indeed, despite its great predictive power and its theoretical appeal, different reasons make one suspect that the Standard Model is an incomplete theory. The author makes a comparative review of the potential for discovering new physics in the upcoming runs 2 and 3 of the LHC, in its possible future High-Luminosity upgrade, and in the two proposed future linear colliders – the Compact Linear Collider (CLIC) and the International Linear Collider (ILC).

This volume ends with an authoritative contribution by Abdelhak DJOUADI, a leading phenomenologist at the Laboratoire de Physique Théorique at Université d'Orsay (now part of the new Université Paris-Saclay). His article, entitled "Implications of the H Boson Discovery", complements and extends that by P. Fayet, summarizing the implications of the discovery of a scalar boson with a mass of approximately 125 GeV in the context of both the Standard Model and its minimal supersymmetric extension, called MSSM. Djouadi first describes how the new experimental data both confirm the H boson as the cornerstone of the Standard Model and imply strong new constraints on particle physics. Fundamental consequences may reach as far as the Universe's fate, ultimately linked to the stability of the H boson potential. This thorough review continues with an even more exciting foray into the consequences for physics beyond the Standard Model. The main scenarios based on supersymmetry are discussed in detail and confronted to experimental data. The conclusion stresses that although supersymmetry has not been observed, it is also not ruled out and still remains the most natural protection of the Higgs mass against radiative corrections, and a road towards a possible grand-unification of electroweak and strong interactions. The next round of data from the LHC is eagerly expected.

This book, by the breadth of topics covered in the celebrated prediction and discovery of the H boson, should be of broad interest to physicists, mathematicians, and historians of science. We further hope that the continued publication of this series of Proceedings will serve the scientific community, at both the graduate and professional levels. We thank the COMMISSARIAT À L'ÉNERGIE ATOMIQUE ET AUX ÉNERGIES ALTERNATIVES (Direction de la Recherche Fondamentale), the DANIEL IAGOLNITZER FOUNDATION, and the ÉCOLE POLYTECHNIQUE for sponsoring this Seminar, and the INSTITUT HENRI POINCARÉ in Paris for its continuing support and hospitality. Special thanks are due to Chantal DELONGEAS for the preparation of the manuscript and to Lars BRINK and David A. KOSOWER for their helpful remarks.

Paris, Saclay   COSTAS BACHAS   BERTRAND DUPLANTIER
& Orsay   Laboratoire de   Institut de
October 2016   Physique Théorique   Physique Théorique
                    de l'ENS   CEA, CNRS
                    École Normale Supérieure   Université Paris-Saclay
                    Paris, France   Gif-sur-Yvette, France
                    bachas@lpt.ens.fr   bertrand.duplantier@cea.fr

                    VINCENT RIVASSEAU
                    Laboratoire de
                    Physique Théorique
                    Université Paris-Saclay
                    Orsay, France
                    rivass@th.u-psud.fr

The H Boson, 1–19
© 2017 Springer Basel AG

The BEH Mechanism and its Scalar Boson

François Englert

Nobel Lecture delivered by Professor François Englert on 8 December 2013, at Aula Magna, Stockholm University. He was introduced by Professor Lars Brink, Chairman of the Nobel Committee for Physics. (See `http://www.nobelprize.org/nobel_prizes/physics/laureates/2013/englert-lecture.html`.)

1. Introduction: short and long range interactions

Physics, as it is conceived today, attempts to interpret the diverse phenomena as particular manifestation of testable general laws. Since its inception at the Renaissance, mainly by Galileo's revolutionary concepts, this has been an extraordinary successful adventure. To the point that after impressive developments in the first half of the twentieth century, one might have even conceived that all phenomena, from the atomic scale to the edge of the visible universe, be governed solely by two fundamental laws, and two known laws. Namely classical general relativity, Einstein's generalisation of Newtonian gravity, and quantum electrodynamics, the quantum version of Maxwell's electromagnetic theory.

Gravitational and electromagnetic interactions are *long range* interactions, meaning they act on objects no matter how far they are separated from each other. The progress in the understanding of such physics applicable to large scales is certainly tributary to the fact they can be perceived without the mediation of highly sophisticated technical devices. But the discovery of subatomic structures had revealed the existence of other fundamental interactions that are *short range*, that is negligible at larger distance scales. In the beginning of the 60s, there was no consistent theoretical interpretation of short range fundamental interactions, neither of the "weak interactions" responsable for radioactive decays, nor of the "strong interactions" responsable for the formation of nuclear structures.

Robert Brout and I [1], and independently Peter Higgs [2], constructed a mechanism to describe short range fundamental interactions. Robert Brout passed away in 2011 and left me alone to tell our story. I will explain how we were led to propose the mechanism, how it allows for consistent fundamental theories of short range interactions and for building elementary particle masses. It became a cornerstone of the Standard Model and was recently confirmed by the magnificent discovery at CERN of its predicted scalar boson.

We became convinced that a consistent formulation of short range interactions would require a common origin for both short and long range interactions.

While both classical general relativity and quantum electrodynamics describe long range interactions and are both built upon very large symmetries, labeled "local symmetries", they have very different structures: in contradistinction to general relativity, the long range quantum electrodynamics is fully consistent at the quantum level and was experimentally verified at that level, in particular by the successful inclusion of chemistry in the realm of known physics. As a valid theory of short range interactions clearly required quantum consistency, we were naturally driven to take, as a model of the corresponding long range interactions, the generalisation of quantum electrodynamics, known as Yang–Mills theory.

The quantum constituents of electromagnetic waves are "photons", massless neutral particles traveling with the velocity of light. Their massless character implies that the corresponding waves are polarised only in directions perpendicular to their propagation. These features are apparently protected by the local symmetry, as the latter does not survive the explicit inclusion of a mass term in the theory. Yang–Mills theory is built upon similar local symmetries, enlarged to include several massless interacting quantum constituents, neutral and charged ones. These massless objects are labelled gauge vector bosons (or often simply gauge bosons).

To transmute long range interactions into short range ones in the context of Yang–Mills theory it would suffice to give these generalised photons a mass, a feature that, as we just indicated, is apparently forbidden by the local symmetries. Leaving momentarily aside this feature, let us first recall why massive particles transmit in general short-range interactions.

Figure 1 is a Feynman diagram whose intuitive appearance hides a precise mathematical content. Viewing time as running from bottom to top, it describes the scattering of two electrons resulting from the exchange of a massive particle labeled Z of mass m_Z. Classically such process could not occur as the presence of the Z particle would violate energy conservation. Quantum mechanically it is allowed if the violation takes place within a time span of the order $\hbar/mc^2$. This process then describes in lowest order perturbation theory a short-range interaction cut-off at a range $\sim \hbar/mc$.

As local symmetries apparently prevent the introduction of massive gauge bosons in the theory, we turn our attention to a class of theories where the state of a system is asymmetric with respect to the symmetry principles that govern its dynamics. This is often the case in the statistical physics of phase transitions [3]. This is not surprising since more often than not energetic considerations dictate that the ground state or low lying excited states of a many body system become ordered. A collective variable such as magnetisation picks up expectation value, which defines an order parameter that otherwise would vanish by virtue of the symmetry encoded in the formulation of the theory (isotropy in the aforementioned example). This is an example of Spontaneous Symmetry Breaking (SSB) which

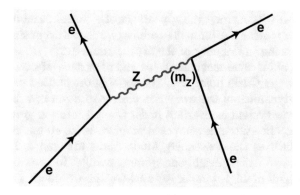

FIGURE 1. Massive particle mediating short-range interactions.

frequently occurs in the statistical theory of second-order phase transitions. Could mass of gauge bosons arise through a similar SSB? This question arises naturally from the seminal work of Nambu who showed that SSB could be transferred from the statistical theory of phase transitions to the realm of relativistic quantum field theory [4, 5, 6], the mathematical framework designed to analyse the world of elementary particles.

This raises a deeper question: could SSB be the agent of the transmutation of long rang interactions mediated by massless gauge fields to short range interactions mediating by massive ones without impairing the validity of the quantum behaviour that characterise the simplest Yang–Mills theory, namely quantum electrodynamics?

As we shall see the answer is yes to both questions *provided the notion of SSB is traded for a more subtle one: the BEH mechanism* [1, 2]. To prepare for the discussion of the mechanism, I will first review how SSB can be transferred from the theory of phase transitions to relativistic quantum field theory.

2. Spontaneous symmetry breaking

2.1. Spontaneous symmetry breaking in phase transitions

Consider a condensed matter system, whose dynamics is invariant under a continuous symmetry. As the temperature is lowered below a critical one, the symmetry may be reduced by the appearance of an ordered phase. The breakdown of the original symmetry is always a discontinuous event at the phase transition point but the order parameters may set in continuously as a function of temperature. In the latter case the phase transition is second order. Symmetry breaking by a second-order phase transition occurs in particular in ferromagnetism, superfluidity and superconductivity.

I first discuss the ferromagnetic phase transition which illustrates three general features of the SSB which set in at the transition point in the low temperature phase: ground state degeneracy, the appearance of a "massless mode" when the dynamics is invariant under a *continuous* symmetry, and the occurrence of a "massive mode" characterising the rigidity of the order parameter.

In absence of external magnetic fields and of surface effects, a ferromagnetic substance below the Curie point displays a global orientation of the magnetisation, while the dynamics of the system is clearly rotation invariant, namely the Hamiltonian of the system is invariant under the full rotation group. This is SSB.

A ferromagnetic system is composed of microscopic atomic magnets (in simplified models such as the Heisenberg Model these are spin 1/2 objects) whose interactions tend to orient neighbouring ones parallel to each other. No global orientation appears at high temperature where the disordering thermal motion dominates. Below a critical "Curie temperature" energy considerations dominate and the system picks up a global magnetisation. The parallel orientation of neighbouring magnets propagates, ending up in a macroscopic magnetisation. This selects a direction, which for an infinite isolated ferromagnet is arbitrary. It is easily proven that for an infinite system any pair of possible orientations define orthogonal ground states and any local excitations on top of these ground states are also orthogonal to each other. Thus the full Hilbert space of the system becomes split into an infinity of disjoint Hilbert spaces. This is ground state degeneracy.

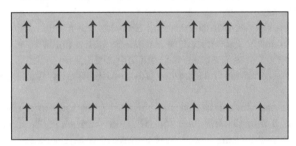

FIGURE 2. The ground state of a ferromagnet.

The effective thermodynamical potential V whose minimum yields the magnetisation in absence of external magnetic field is depicted in Figure 2. Above the Curie point T_c the magnetisation $\vec{M}$ vanishes. Below the Curie point in a plane $V M_z$ the potential develops in a plane $V M_z$ a double minimum which generate a valley in the M_x, M_y directions. Each point of the valley defines one of the degenerate ground states with the same $|\vec{M}|$.

At a given minimum, say, $\vec{M} = M^z \vec{1}_z$, the curvature of the effective potential measures the inverse susceptibility which determines the energy for infinite wavelength fluctuations. This is the analog of mass in relativistic particle physics. The inverse susceptibility is zero in directions transverse to the order parameter and positive in the longitudinal direction. One thus obtains from the transverse susceptibility a "massless" transverse mode characteristic of broken continuous symmetry: these are the "spin-waves" whose quantum constituents are interacting

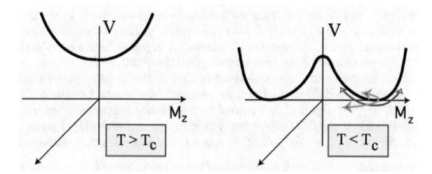

FIGURE 3. Effective thermodynamical potential of a ferromagnet above and below the Curie point.

bosons called "magnons". The longitudinal susceptibility yields a (possibly unstable) "massive" longitudinal mode which corresponds to fluctuations of the order parameter. In contradistinction to the massless mode which exists only in continuous SSB for which there is a valley, the massive mode is present in any SSB, continuous or discrete and measures the rigidity of the ordered structure.

The structure of Figure 3 is common to many second-order phase transitions and leads to similar consequences. However in superconductivity a new phenomenon occurs. The quantum phase symmetry is broken by a condensation of electron pairs bounded by an attractive force due to phonon exchange in the vicinity of the Fermi surface. The condensation leads to a energy gap at the Fermi surface. For neutral superconductors, this gap would host a massless mode and one would recover the general features of SSB. But the presence of the long-range Coulomb interactions modifies the picture. *The massless mode disappears*: it is absorbed in electron density oscillations, namely in the "massive" plasma mode. As will be apparent later, this is a precursor of the BEH mechanism [7, 4, 8].

2.2. Spontaneous symmetry breaking in field theory

Spontaneous symmetry breaking was introduced in relativistic quantum field theory by Nambu in analogy with the BCS theory of superconductivity [4]. The problem studied by Nambu [5] and Nambu and Jona-Lasinio [6] is the spontaneous breaking of the $U(1)$ symmetry of massless fermions resulting from the arbitrary relative (chiral) phase between their decoupled right and left constituent neutrinos. Chiral invariant interactions cannot generate a fermion mass in perturbation theory but may do so from a (non-perturbative) fermion condensate: the condensate breaks the chiral symmetry spontaneously. Nambu [5] showed that such spontaneous symmetry breaking is accompanied by a massless pseudoscalar. This is interpreted as the chiral limit of the (tiny on the hadron scale) pion mass. Such interpretation of the pion constituted a breakthrough in our understanding of strong interaction physics. The massless pseudoscalar is the field-theoretic counterpart of the "massless" spin-wave mode in ferromagnetism. In the model of reference [6], it is shown that SSB also generates a massive scalar boson which is the counterpart of the "massive mode" measuring in phase transitions the rigidity of the order parameter in the spontaneously broken phase.

The significance of the massless boson and of the massive scalar boson occurring in SSB is well illustrated in a simple model devised by Goldstone [9]. The potential $V(\phi_1, \phi_2)$ depicted in Figure 4 has a rotational symmetry in the plane of the real fields (ϕ_1, ϕ_2), or equivalently is invariant under the $U(1)$ phase of the complex field $\phi = (\phi_1 + i\phi_2)/\sqrt{2}$. This symmetry is spontaneously broken by the

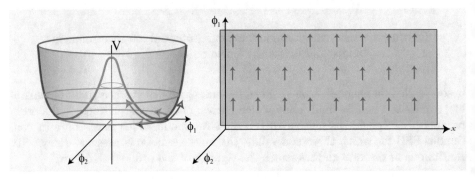

FIGURE 4. Spontaneous symmetry breaking in the Goldstone model.

expectation value $\langle\phi\rangle$ of the ϕ-field acquired at a minimum of the potential in some direction of the (ϕ_1, ϕ_2) plane, say $\langle\phi_1\rangle$. Writing $\phi = \langle\phi\rangle + \varphi$

$$\phi_1 = \langle\phi_1\rangle + \varphi_1, \tag{2.1}$$

$$\phi_2 = \varphi_2. \tag{2.2}$$

For small φ_1 and φ_2 we may identify the quantum fluctuation φ_1 climbing the potential as the massive mode measuring the rigidity of the SSB ground state selected by $\langle\phi_1\rangle$, and the quantum fluctuation φ_2 in the orthogonal valley direction as the massless mode characteristic of a continuous SSB.

Their significance is illustrated in Figures 5 and 6 depicting respectively classical φ_2 and φ_1 wave modes, on the classical background $\langle\phi_1\rangle$. The corresponding massless and massive bosons are the quantum constituents of these waves.

Figure 5a represents schematically a lowest energy state (a "vacuum") of the system: a constant non-zero value of the field $\phi_1 = \langle\phi_1\rangle$ pervades space-time. Figure 5b depicts the excitation resulting from the rotation of half the fields in the (ϕ_1, ϕ_2) plane. This costs only an energy localised near the surface separating the rotated fields from the chosen vacuum. SSB implies indeed that rotating all the fields would cost no energy at all: one would merely trade the initial chosen vacuum for an equivalent one with the same energy. This is the characteristic *vacuum degeneracy* of SSB. Figure 5c mimics a wave of φ_2. Comparing 5c with 5b, we see that as the wavelength of the wave increases indefinitely, its energy tends to zero,

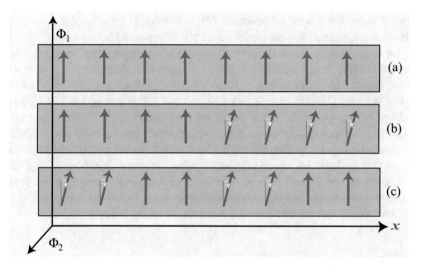

FIGURE 5. Massless Nambu–Goldstone mode φ_2.

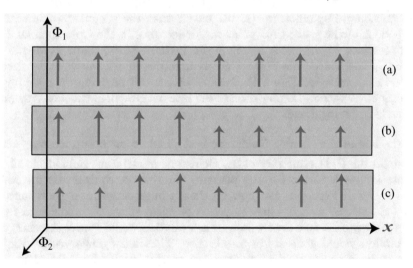

FIGURE 6. Massive scalar mode φ_1.

and may be viewed as generating in that limit a motion along the valley of Figure 4. Quantum excitations carried by the wave reach thus zero energy at zero momentum and the mass m_{φ_2} is zero. Figure 5 can easily be generalised to more complex spontaneous symmetry breaking of *continuous* symmetries. Massless bosons are thus a general feature of such SSB already revealed by Nambu's discovery of the massless pion resulting from spontaneous chiral symmetry breaking [5]. They will be labeled *massless Nambu–Goldstone (NG) bosons*. Formal proofs corroborating the above simple analysis can be found in the literature [10].

Figure 6 depicts similarly a classical wave corresponding to a stretching of the vacuum fields. These excitations in the ϕ_1 direction describe fluctuations of the order parameter $\langle \phi_1 \rangle$. They are volume effects and their energy does not vanish when the wavelength becomes increasingly large. They correspond in Figure 4 to a climbing of the potential. The quantum excitations φ_1 are thus now massive. These considerations can be again extended to more general SSB (even to discrete ones) to account for order parameter fluctuations. Lorentz invariance imposes that such massive excitations are necessarily scalar particles. They were also already present in reference [6] and will be denoted in general as *massive scalar bosons*.

To summarise, φ_2 describes massless bosons, φ_1 massive ones, and the "order parameter" $\langle \phi_1 \rangle$ may be viewed as a condensate of φ_1 bosons.

3. The BEH mechanism

The above considerations are restricted to spontaneous symmetry breaking of *global* continuous symmetries. Global means that the symmetry operations are independent of the space-time point x. For instance in the Goldstone model the *global* rotations of the fields in Figure 5a in the (ϕ_1, ϕ_2) plane by angles independent of the space-time point x are symmetries of the theory (they describes motion in the valley of Figure 4): these rotations cost no energy and simply span the degenerate vacua. We now discuss the fate of SSB when the global symmetry is extended to a local one.

3.1. The fate of the Nambu–Goldstone boson and vector boson masses

We extend the $U(1)$ symmetry of the Goldstone model from global to local. Thus the rotation angle in the (ϕ_1, ϕ_2) plane in Figure 5, or equivalently the rotation in the valley of Figure 4, can now be chosen independently at each space-time point (x) with no cost of energy and no physical effect. To allow such feature, one has to invent a new field whose transformation would cancel the energy that such motion would generate in its absence. This is a "gauge vector field" A_μ. It has to be a vector field to compensate energy in all space directions and it has to transform in a definite way under a rotation in the (ϕ_1, ϕ_2) plane: this is called a gauge transformation and results in a large arbitrariness in the choice of the A_μ field corresponding to arbitrary "internal" rotations at different points of space. The consequence of this gauge symmetry is that the waves are polarised in directions perpendicular to their direction of propagation and that there quantum constituents have to be introduced as massless objects.

Local $U(1)$ symmetry is the simplest gauge field theory and is the symmetry group of quantum electrodynamics. In the local generalisation (the gauging) of the Goldstone model, the introduction of the potential of Figure 4 will deeply affect the "electromagnetic potential" A_μ.

As in the Goldstone model of Section 1.2, the SSB Yang–Mills phase is realized by a non-vanishing expectation value for $\phi = (\phi_1 + i\phi_2)/\sqrt{2}$, which we choose to be in the ϕ_1-direction. Thus

$$\phi = \langle \phi \rangle + \varphi, \tag{3.1}$$

with $\phi_1 = \langle \phi_1 \rangle + \varphi_1$ and $\phi_2 = \varphi_2$. As previously φ_2 and φ_1 appear to describe a NG massless boson and a massive scalar boson.

However a glance on Figure 5 depicting the NG mode immediately shows that Figure 5b and Figure 5c differs from Figure 5a only by local rotations and hence in the local Goldstone model, they are just symmetry (or equivalently gauge) transformations. They cost no energy and therefore the NG boson has disappeared: the corresponding fluctuations in the valley are just redundant (gauge transformed) description of the same gauge invariant vacuum. It is easy to see that this argument remains valid for any local symmetry and hence *Nambu–Goldstone bosons do not survive the gauging of a global SSB to a local symmetry*. The vacuum is no more degenerate and strictly speaking there is no spontaneous symmetry breaking of a local symmetry. The reason why the phase with non-vanishing scalar expectation value is often labelled SSB is that one uses perturbation theory to select at zero coupling with the gauge fields a scalar field configuration from global SSB; but this *preferred* choice is only a convenient one.

The disappearance of the NG boson is thus an immediate consequence of local symmetry. The above argument [11] was formalized much later [12] but formal proofs not directly based on the gauge invariance of the vacuum were already presented in 1964 [13, 14].

One may now understand in qualitative terms the consequence of the disappearance of the NG boson. Clearly, one does not expect that the degrees of freedom carried by the NG ϕ_2 field could vanish. As the NG boson disappears because of its coupling to the gauge field, one expects that these degrees of freedom should be transferred to it. This can only occur by adding to the transverse polarisation of the gauge field a longitudinal one. But such polarisation is forbidden as mentioned earlier, for a massless field. Therefore the coupling of the would be NG boson to the gauge field must render the latter massive! This is the essence of the BEH mechanism.

These qualitative considerations can be made quantitative [1] by considering the Feynman graphs (time runs horizontally) describing the propagation of the A_μ gauge field in the vacuum with non-vanishing scalar field expectation value, say $\langle \phi \rangle \neq 0$. This propagation is depicted in lowest order in Figure 7 (time runs horizontally) and the interaction of A_μ with the condensate $\langle \phi \rangle$ amounts to a "polarisation" of the vacuum. The first graph shows the local interaction of the gauge field with the condensate while the second one gives a non-local interaction due to the propagation of a NG boson. Here e is the coupling of the gauge vector to matter, q_μ is a four-momentum (q_0 =energy; $\vec{q}$ = momentum), $q^2 = q_0^2 - \vec{q}^2$

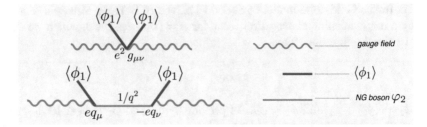

FIGURE 7. Interaction of the gauge field with the condensate.

and $g_{\mu\nu}$ has only non-zero values if $\mu \neq \nu : 1, -1, -1, -1$. The two graphs sum to

$$\Pi_{\mu\nu} = (g_{\mu\nu} - q_\mu q_\nu/q^2)\Pi(q^2), \qquad (3.2)$$

where

$$\Pi(q^2) = e^2\langle\phi_1\rangle^2. \qquad (3.3)$$

The second factor of Eq. (3.2) does not vanish when $q^2 = 0$. In field theory this means that the gauge field has acquired a mass

$$M_V^2 = e^2\langle\phi_1\rangle^2. \qquad (3.4)$$

The first factor describes the projection at $q^2 = M_V^2$ of $g_{\mu\nu}$ on a *three*-dimensional space of polarisations, which, as explained in qualitative terms above, is required for a massive vector. Its transversality (i.e., its vanishing under multiplication by q^μ) is characteristic of a "Ward Identity" which expresses that the local gauge symmetry has not been broken and is identical to the analogous factor in quantum electrodynamics, an important fact that will be commented upon in the following section.

We now discuss the generalisation of these results to more complicated symmetries. One gets (for real fields) a mass matrix

$$(M_V^2)^{ab} = -e^2\langle\phi^B\rangle T^{a\,BC}T^{b\,CA}\langle\phi^A\rangle, \qquad (3.5)$$

where $T^{a\,BC}$ is a real anti-symmetric generator coupled to a gauge field A_μ^a and $\langle\phi^A\rangle$ designates a non-vanishing expectation value.

In these cases, some gauge fields may remain massless. Consider for instance instead of the invariance of the Goldstone model on a circle in the plane (ϕ_1, ϕ_2), an invariance on a sphere in a three-dimensional space (ϕ_1, ϕ_2, ϕ_3) broken by $\langle\phi_1\rangle \neq 0$. There are now three gauge fields associated to the rotations on the sphere, and while A_μ^2 and A_μ^3 acquire mass A_μ^1 remains massless. This can be understood in the following way: rotation generators around the directions 2 and 3 would move $\langle\phi_1\rangle$ if the symmetry were global and would thus give rise, as in Figure 5, to NG bosons; their degrees of freedom are transferred in local symmetries to the massive gauge vector fields A_μ^2 and A_μ^3, providing their third degree of polarisation. The expectation value $\langle\phi_1\rangle$ is not affected by rotation generators around the direction

1 and does not generate in the global symmetry case NG bosons and hence the corresponding A_μ^1 remains massless.

Thus the BEH mechanism can unify in the same theory long and short range interactions by leaving unbroken a subgroup of symmetry transformation (e.g., rotation around the direction 1) whose corresponding gauge fields remain massless.

3.2. The fate of the massive scalar boson

A glance at Figure 5 shows that the stretching of (classical) scalar fields is independent of local rotations of the ϕ-field in the (ϕ_1, ϕ_2) plane. This translates the fact that the modulus of the ϕ-field is gauge invariant. Hence the scalar bosons survives the gauging and their classical analysis is identical to the one given for the Goldstone model in Section 1.2. The coupling of the scalar boson φ_1 to the massive

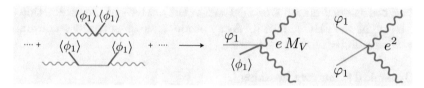

FIGURE 8. Coupling of the scalar boson φ_1 to massive gauge bosons.

gauge bosons follows from the Figure 7, by viewing the Feynman diagrams with time going from top to bottom and using Eq. (3.1). One gets the two vertices of Figure 8 where the heavy wiggly lines on the right-hand side represent the massive gauge propagators. The vertex couplings follow from Eq. (3.4).

3.3. Fermion masses

Let us couple the Yang–Mills fields to massless fermions in a way that respects the Yang–Mills symmetry. This coupling preserves the chiral symmetry of the massless fermions and fermion mass requires SSB. In the Nambu theory of spontaneous breaking of chiral symmetry, this gives rise to NG bosons which are here eaten up by massive gauge fields. This can be done by suitable couplings of the scalar fields whose expectation value breaks the symmetry. Mass generation for fermions is depicted in Figure 9.

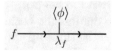

FIGURE 9. Mass generation $m_f = \lambda_f \langle \phi \rangle$ from a coupling λ_f of fermions to the scalar field ϕ.

3.4. Why is the mechanism needed?

Eq. (3.2) expresses that the mass generation from the BEH mechanism does not destroy the local symmetry, in contradistinction to a mass term introduced by hand ab initio. This equation remains valid at higher orders in perturbation theory and has the same form as the polarisation in quantum electrodynamics. As in the latter case, it implies that in covariant gauges, the gauge vector boson propagator tames the quantum fluctuations, and suggest therefore that the theory is renormalizable [15]. However it is a highly non trivial matter to prove that it does not introduce contributions from unphysical particles and it is therefore a very difficult problem to prove quantum consistency to all orders. That this is indeed the case has been proven by 't Hooft and Veltman [16] (see also Ref. [17]).

The quantum consistency of the BEH mechanism is the basic reason for its success. Precision experiments can be predicted and were indeed verified. The quantum consistency played a critical rôle in the analysis of the production of the scalar boson at the LHC and of its decay products, leading to the confirmation of the detailed validity of the mechanism.

3.5. Dynamical symmetry breaking

The symmetry breaking giving mass to gauge vector bosons may also arise from a fermion condensate. This is labeled *dynamical symmetry breaking*. If a spontaneously broken global symmetry is extended to a local one by introducing gauge fields, the massless NG bosons disappear as previously from the physical spectrum and their absorption by gauge fields renders these massive. In contradistinction with breaking by scalar field condensate, it is very difficult in this way to give mass in a renormalizable theory simultaneously to both gauge vector fields and fermions.

3.6. The electroweak theory and the Standard Model

The most impressive success of the BEH mechanism is the electroweak theory for weak and electromagnetic interactions [18] applied to all particles of the Standard Model. These encompasses all known particles. These are a) the fermions which are listed in Figure 10, b) γ and W^+, W^-, Z, the gauge vector bosons transmitting the electromagnetic and the weak interactions, c) eight "gluons", the gauge vectors bosons of the "color group" $SU(3)$ mediating the strong interactions, and d) last but not least, one massive scalar boson which was recently discovered and identified as the scalar predicted by the BEH mechanism.

The first row in Figure 10 contains the basic constituents of the proton and the neutron, namely the electron, the three up and down coloured quarks building the proton and neutron bonded by the gluons, to which is added the electron neutrino. The second [19] and third row [20] were completed as predictions in the seventies and verified afterwards. Color was also introduced in the seventies. The particles in the first and the second row are called leptons. To all fermions of the table, one must of course also add their antiparticles.

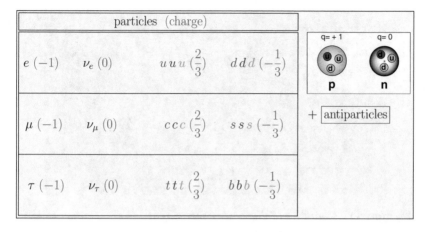

FIGURE 10. Fermion constituents of the Standard Model.

All the fermions are chiral and their chiral components have different group quantum numbers. Hence they are, as the gauge vector bosons, massless in absence of the BEH mechanism, i.e., in absence of the scalar condensate. The condensate $\langle\phi\rangle \neq 0$ gives mass to the W^+, W^-, Z bosons and to all fermions except the three chiral neutrinos which have no opposite chirality counterpart in the conventional Standard Model. The photons and the gluons remain massless but the latter become short range due in the conventional description to a highly non-perturbative vacuum (resulting from a mechanism somehow dual to the BEH mechanism).

The discovery of the Z and W bosons in 1983 and the precision experiments testing the quantum consistency of the Standard Model established the validity of the mechanism, but it was still unclear whether this was the result of a dynamical symmetry breaking or of a particle identifiable as a elementary boson at the energy scale considered.

4. The discovery

In the Standard Model, there is one real massive scalar boson φ (also labelled H). It couples to the massive W and Z bosons. This follows from Figure 8 and the couplings are depicted in Figure 11a. Its coupling to elementary fermions similarly follows from the couplings in Figure 9 as shown in Figure 11b. The coupling to the massless photons is a genuine quantum effect involving loops, even in lowest order, as indicated in Figure 12. The LHC site circling under the French-Swiss border is schematically indicated in the picture of Figure 13. The 27 km circular tunnel containing two opposite beams of protons surrounded by guiding superconducting electromagnets cooled by superfluid helium is pictured in Figures 14, 15 and 16 are pictures of the ATLAS and CMS detectors at diametrical opposite sites of the tunnel. There collisions occur and were used primarily to detect and identify the

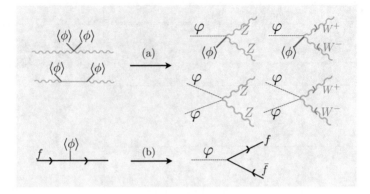

FIGURE 11. Coupling of the scalar boson φ to massive gauge bosons and to elementary fermions.

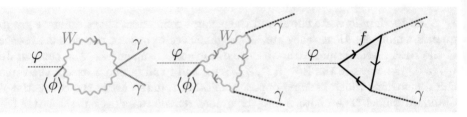

FIGURE 12. Coupling of the scalar boson φ to photons.

scalar boson of the Standard Model (and possibly other ones). At the end of 2012, proton-proton collisions occurred at the rate of nearly 10^9 sec^{-1} and the proton energy reached 8 TeV. At these energies, all quarks of Figure 10 and the gluons connecting them may contribute to the production of the scalar boson. The leading production processes are represented in Figure 17.

As an example of the data gathered by CMS and ATLAS, Figure 18 gives the data obtained by the CMS group of observed decays into 4 leptons at the end the 2012 run. The blue area is the expected background, namely those decays which would follow from the Standard Model if, at given total mass, there would be no contribution from the scalar boson. The red curve measures the contribution that could be due to the scalar decaying into two Z vector bosons which further decay into leptons, as was confirmed by further analysis. Note that one of the Z is real but the other is "virtual", meaning that this decay is forbidden by energy conservation but may contribute in the quantum theory. Consideration of other decay channels and spin analysis show that the particle detected is consistent with the Standard Model scalar with a mass $m_H \simeq 125$ GeV. The absence of new particles at comparable energies, as well as the success of the Feynman graph analysis including loops, points towards an elementary particle, at least up to

FIGURE 13. Schematic location of the LHC.

FIGURE 14. The LHC dipole magnets.

the energy range considered. This is the first elementary spin zero particle ever detected. It raises the interesting possibility of supersymmetry broken at attainable energies, although there are no indication of it so far.

The elementary character of the scalar already eliminates many dynamical models of symmetry breaking and raises interesting possibilities for extrapolation

FIGURE 15. The Atlas detector.

FIGURE 16. The CMS detector.

beyond presently known energies, up to those close to the Planck scale where quantum gravity effects might play a dominant rôle. The analysis of these speculations is beyond the scope of this talk.

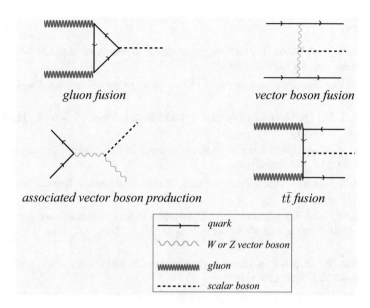

FIGURE 17. Production of the Standard Model scalar boson.

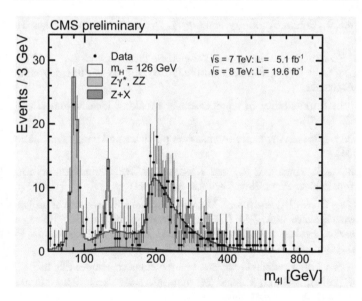

FIGURE 18. Decay of the scalar boson into 4 leptons from two Z's.

References

[1] Englert, F., and Brout, R.: Broken symmetry and the mass of gauge vector mesons. Phys. Rev. Letters **13**, 321 (1964).

[2] Higgs, P.W.: Broken symmetries and the masses of gauge bosons. Phys. Rev. Lett. **13** 508 (1964).

[3] Landau, L.D.: On the theory of phase transitions I. Phys. Z. Sowjet. **11**, 26 (1937), [JETP **7**, 19 (1937)].

[4] Nambu, Y.: Quasi-particles and gauge invariance in the theory of superconductivity. Phys. Rev. **117**, 648 (1960).

[5] Nambu, Y.: Axial vector current conservation in weak interactions. Phys. Rev. Lett. **4**, 380 (1960).

[6] Nambu, Y., and Jona-Lasinio, G.: Dynamical model of elementary particles based on an analogy with superconductivity I, II. Phys. Rev. **122**, 345 (1961); **124**, 246 (1961).

[7] Anderson, P.W.: Random-phase approximation in the theory of superconductivity. Phys. Rev. **112**, 1900 (1958).

[8] Anderson, P.W.: Plasmons, gauge invariance, and mass. Phys. Rev. **130**, 439 (1963).

[9] Goldstone, J.: Field theories with "superconductor" solutions, Il. Nuovo Cimento **19**, 154 (1961).

[10] Goldstone, J., Salam, A., and Weinberg, S.: Broken symmetries. Phys. Rev. **127**, 965 (1962).

[11] Englert, F.: *Broken symmetry and Yang–Mills theory*. In 50 *years of Yang–Mills theory*, ed. by G. 't Hooft, World Scientific (2005), pp. 65–95, http://arxiv.org/abs/hep-th/0406162.

[12] Elitzur, S.: Impossibility of spontaneously breaking local symmetries. Phys. Rev. **D12**, 3978 (1975).

[13] Higgs, P.W.: Broken symmetries, massless particles and gauge fields. Phys. Lett. **12**, 132 (1964).

[14] Guralnik, G.S., Hagen, C.R., and Kibble, T.W.B.: Global conservation laws and massless particles. Phys. Rev. Lett. **13**, 585 (1964).

[15] Englert, F., Brout, R., and Thiry, M.: Vector mesons in presence of broken symmetry, Il. Nuovo Cimento **43A**, 244 (1966); Proceedings of the 1997 Solvay Conference, *Fundamental Problems in Elementary Particle Physics*, Interscience Publishers J. Wiley and Sons, p. 18.

[16] 't Hooft, G.: Renormalizable lagrangians for massive Yang–Mills fields. Nucl. Phys. **B35**, 167 (1971); 't Hooft, G., and Veltman, M.: Regularization and renormalization of gauge fields. Nucl. Phys. **B44**, 189 (1972).

[17] Lee, B.W., and Zinn-Justin, J.: Spontaneously broken gauge symmetries. Phys. Rev. **D5**, 3121; 3137; 3155 (1972).

[18] Glashow, S.L.: Partial-symmetries of weak interactions. Nucl. Phys. **22**, 579 (1961); Weinberg, S.: A model of leptons. Phys. Rev. Lett. **19**, 1264 (1967); Salam, A.: Proceedings of the 8th Nobel Symposium, *Elementary Particle Physics*, ed. by N. Svartholm, (Almqvist and Wiksell, Stockhlom) p. 367.

[19] Glashow, S.L., Iliopoulos, J., and Maiani, L.: Weak interactions with lepton-hadron symmetry. Phys. Rev. **D2**, 1285 (1970).

[20] Kobayashi, M., and Maskawa, T.: CP Violation in the renormalizable theory of weak interaction. Prog. Theor. Phys. **49**, 652 (1973).

François Englert
Service de Physique Théorique
CP225
Université Libre de Bruxelles
Boulevard du Triomphe (Campus de la Plaine)
B-1050 Bruxelles, Belgium
e-mail: `englert@ulb.ac.be`

The H Boson, 21–50
© 2017 Springer Basel AG

Discovery and Measurements of the H Boson with ATLAS and CMS Experiments at the LHC

Yves Sirois*

Abstract. The discovery of a new boson with a mass around 125 GeV has been established in July 2012 by the ATLAS and CMS experiments at the LHC collider. More than twice as much data was collected by the end of 2012. The analysis of the full data sample, collected with pp collisions at 7 and 8 TeV in 2011 and 2012, has now allowed for considerable progress in understanding the nature of the new boson.

The new boson is found to be a Higgs boson, with properties as expected for the scalar boson H resulting from the Brout–Englert–Higgs mechanism responsible for electroweak symmetry breaking in the standard model. A review of the latest ATLAS and CMS results on the H boson is presented here.

1. Introduction: the standard model and the Higgs boson

Over the past four decades, the standard model (SM) of particle physics has provided a remarkably accurate description of numerous results from accelerator and non-accelerator based experiments. Yet, until very recently, the question of how the W and Z gauge bosons acquire mass remained an opened question. This question could have jeopardized the validity of the theory at higher energies or, equivalently, at smaller distance scales. Understanding the origin of the electroweak symmetry breaking (EWSB), how the W and Z bosons acquire mass whilst the photon remains massless, has been set as one of the most important objectives of the Large Hadron Collider (LHC) physics program at the birth of the project more than twenty years ago. The SM remained an unchallenged [1] but incomplete theory

*A similar review is in preparation for the "Comptes Rendues de l'Académie des Sciences" and written in collaboration with, Rosy Nicolaidou, IRFU, CEA-Saclay.

for the interactions of particles until the LHC finally provided its first high energy proton-proton collisions at 7 TeV in 2010. The discovery of a new boson at a mass of about 125 GeV by the ATLAS [2] and CMS [3, 4] experiments in 2012, and the confirmation with additional data that the boson behaves like a Higgs boson, have now considerably changed the landscape.

The SM comprises matter fields, the quarks and leptons as the building blocks of matter, and describes their interactions through the exchange of force carriers: the photon for electromagnetic interactions, the W and Z gauge bosons for weak interactions, and the gluons for strong interactions. The electromagnetic and weak interactions are partially unified in the Glashow–Weinberg–Salam electroweak theory [5, 6, 7]. The gauge bosons are a direct consequence of the underlying gauge symmetries. It is sufficient to postulate the invariance under $SU(2) \times U(1)$ gauge symmetry in the electroweak sector to see emerging as a necessity the existence of the photon, for the electromagnetic interaction, and the W and Z bosons, for the weak interactions. The gauge symmetries are the essential pillars of the theory and thus must be preserved. This is only possible if the gauge bosons remain massless in the fundamental theory. Besides the question of the origin of the mass of vector bosons, the very existence of these massive bosons was threatening the theory at the TeV scale. In contrast to quantum electrodynamics where a renormalizable theory is obtained by injecting the masses and charges measured at a given scale by hand, no such trick is possible for the weak interaction while preserving the gauge symmetries. The massive vector bosons lead to violation of unitarity for calculations at the TeV scale, unless something else is added. The SM with the gauge bosons and matter fields is incomplete. Additional structure is needed.

Since the advent of the electroweak theory, the Brout–Englert–Higgs (BEH) mechanism [8, 9, 10, 11, 12, 13] had been adopted as a solution to both the EWSB and the unitarization of the theory. In this mechanism, the introduction of a complex scalar doublet field with self interactions allows for a spontaneous EWSB. This leads to the generation of the W and Z masses, the weak boson acquiring longitudinal degrees of freedom, and to the prediction of the existence of one physical Higgs boson (H). The fundamental fermions also acquire mass through Yukawa interactions with the scalar field when propagating in the physical vacuum: the left- and right-handed chiralities become coupled. The mass m_H of the Higgs boson in the SM is not predicted by the theory, but general considerations [14, 15, 16, 17] on the finite self-coupling of the Higgs field, the stability of the vacuum, and unitarization bounds suggest that it should be smaller than about 1 TeV. The existence of a scalar boson is sufficient to allow for an exact unitarization of the theory. But saving the theory has a cost: the arbitrariness m_H (and of the self-couplings) and the fact that the Higgs boson is not a gauge boson. Thus the mass m_H is not protected by any symmetry of the theory. The mass is sensitive to any new scale beyond the SM which could contribute in quantum fluctuations. The theory would have to be fine tuned to maintain m_H at the weak scale.

With these considerations in mind, the scene is set to describe the search, the discovery, and the measurements of the Higgs boson at the LHC. This review is organized as follows. First, we briefly describe the ATLAS and CMS experiments in Section 2. We then focus on the H boson and remind about the relevant phenomenology aspects in Section 3. In Section 4, we recollect the adventure of the search for the Higgs boson at the LEP e^+e^- collider, the Tevatron $p\bar{p}$ colllider, and the LHC pp collider including data collected at $\sqrt{s} = 7\,\mathrm{TeV}$ in 2011. The additional data collected at the LHC at $\sqrt{s} = 8\,\mathrm{TeV}$ lead, in July 2012, to the discovery of the new boson via di-boson channels as reminded in Section 4. We then turn in Section 5 to the measurement of the properties of the Higgs boson using all available LHC data from run I, starting with the high resolution channels and the measurement of the Higgs boson mass, then with constraints on the resonance width, tests and constraints on the spin-parity state, comparisons of the signal rates with SM expectation in various production and decay modes, and finally the coupling constraints and compatibility with SM expectation. We conclude in Section 6 with some elements of prospects for the future data taking at the LHC.

2. The ATLAS and CMS experiments

The ATLAS ("A Toroidal LHC Apparatus") [18] and CMS ("Compact Muon Solenoid") [19] detectors are multi-purpose devices with a cylindrical geometry, and forward-backward symmetry along the beam line. Both experiments have been designed to allow for a good measurement of leptons from low to very high momenta, offer sufficient transverse or longitudinal granularity to provide a high discrimination of isolated leptons against QCD instrumental background, and provide a nearly 4π solid angle coverage for the measurement of hadronic jets and transverse energy flow.

Above all other considerations, the detailed design of the experiments follows from the choice of the main magnets. The CMS experiment has chosen a solenoid which allows for a compact detector. The solenoid provides field lines parallel to the Z (beam) axis so that charged particles trajectories bend in the transverse plane. The excellent momentum resolution required for TeV muons is made possible via a very high magnetic field and a fine grained tracker. The ATLAS experiment has chosen a toroid which imposes a very large volume. The toroid provides field lines which are circles centered on the Z axis, so that muons bend in a plane defined by the beam axis and the muon position. This provides excellent stand alone momentum resolution for TeV muons, but an internal solenoid is needed for the purpose of vertex reconstruction and additional momentum measurements with a fine grained tracker. The experiments were ready to take data in 2008, after about 15 years of research and developments, design and construction.

Three-dimensional representations of the ATLAS and CMS detectors are shown Figure 1. The ATLAS layout comprises a thin superconducting solenoid surrounding inner tracking detectors and three large superconducting toroids supporting a large muon tracker. The inner detectors consist of a silicon pixel device, a silicon microstrip device and a transition radiation tracker, all immersed

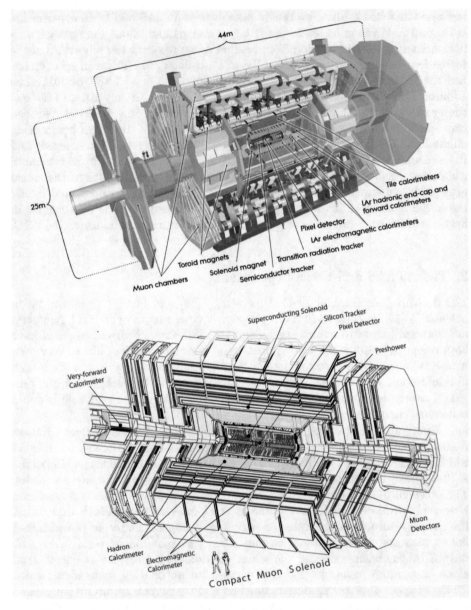

FIGURE 1. Cut-away three-dimensional view of the ATLAS (top) and
CMS (bottom) detectors. The instruments occupy volumes with cylindrical shapes, with dimensions for ATLAS of 44 m in length, 25 m in
diameter and a weight of ∼ 7000 tons, for CMS of 21.6 m in length,
14.6 m in diameter, and 12500 tons.

in the 2 Tesla field from the solenoid. High-granularity liquid-argon (LAr) electromagnetic sampling calorimeters cover the pseudorapidity region $|\eta| < 3.2$. An iron-scintillator tile calorimeter provides coverage over $|\eta| < 1.7$. The end-cap and forward regions, spanning $1.5 < |\eta| < 4.9$, are instrumented with LAr calorimetry for both electromagnetic and hadronic measurements. The muon spectrometer covering $|\eta| < 2.7$ relies on the magnetic deflection of muons tracks in the air-core toroid magnets, instrumented with separate trigger and high-precision tracking chambers The CMS layout comprises a superconducting solenoid of 6 m internal diameter, providing a uniform magnetic field of 3.8 T. The bore of the solenoid is instrumented with various particle detection systems. The inner tracking system is composed of a pixel detector with three barrel layers at radii between 4.4 and 10.2 cm and a silicon strip tracker with 10 barrel detection layers extending outwards to a radius of 1.1m. Each system is completed by two end caps, extending the acceptance up to $|\eta| < 2.5$. A lead tungstate crystal electromagnetic calorimeter with fine transverse ($\Delta\eta, \Delta\phi$) granularity and a brass-scintillator hadronic calorimeter surround the tracking volume and cover the region $|\eta| < 3$. The steel return yoke outside the solenoid is in turn instrumented with gas detectors which are used to identify muons in the range $|\eta| < 2.4$. The barrel region is covered by drift tubes and the end-cap region by cathode strip chambers. A calorimeter made of steel absorber and quartz fiber extends to coverage in forward regions up to $|\eta| < 5.0$.

Both experiments have profited from many months of training and analysis with cosmic data in 2008 and 2009, before the arrival of the first stable LHC collisions. The first LHC collisions were produced at a proton-proton $\sqrt{s} = 900\,\text{GeV}$ during "pilot" runs and then, after a few weeks, at a record $\sqrt{s} = 2.36\,\text{TeV}$. The data collected by the two large experiments, ATLAS and CMS during these early runs were essentially used to finalize the commissioning of the detectors and the analysis tools, and to validate the computing and data distribution models. In December 2009, the level of readiness of the experiments was such, that first physics results could be produced, in some cases within days or weeks. The first public results concerned basic QCD background properties such as the measurement of the underlying event activity, track multiplicity and transverse momentum flow measurements, or for instance the observation of diffraction in proton-proton collisions. After a short technical stop, the LHC operations have resumed in early spring 2010 at $\sqrt{s} = 7\,\text{TeV}$, the highest energy compatible with a secured and stable functioning of the collider. In the following, we concentrate on results obtained at the LHC during the so-called "run I" in 2011 and 2012, with pp collisions at $\sqrt{s} = 7$ and $8\,\text{TeV}$.

3. Phenomenology at the LHC

3.1. Production and decay modes

In pp collisions, the Higgs boson is produced dominantly by a gluon fusion (ggH) process involving a virtual top (or bottom) quark loop. The other main production

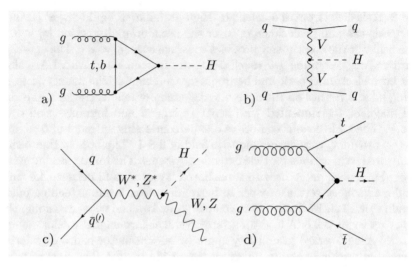

FIGURE 2. Examples of leading order Feynman diagrams contribution
to the production of the SM Higgs boson in hadronic collisions; (a)
gluon-gluon fusion $gg \to H$ through b- and t-quark fermion loops; (b)
vector boson fusion WWH or ZZH; (c) Higgs-strahlung WH or ZH; (d)
associated production of a Higgs boson and a $t\bar{t}$ pair.

modes are the vector boson fusion (VBF), the "Higgsstrahlung" (VH with V=W
or Z), and the associated production ($t\bar{t}$H). The production modes are illustrated
in the Figure 2.

The total production cross sections for a SM Higgs [20] boson at the LHC
are shown as a function of m_H in Figure 3 (left). For $m_H = 125$ GeV, the total
production cross section is of about 22 pb at a centre of mass of $\sqrt{s} = 8$ TeV
(about 17 pb at $\sqrt{s} = 7$ TeV). The Higgs boson is thus expected to be copiously
produced at the LHC. For this mass, about 87% of the Higgs bosons are produced
via ggH, 7.1% via VBF, 4.9% via VH, and 0.6 % via $t\bar{t}$H. It is worth mentioning
that a huge effort to provide the theoretical cross section calculations at next-to-
next-to-leading order (NNLO) level has been done the past years and this effort
continues with increased interest.

The decay branching ratios for a SM Higgs boson [20] are shown in Figure 3
(right). The WW di-boson decay dominates at high masses, for $m_H > 135$ GeV.
The WW and ZZ di-boson decays are the sole relevant modes for $m_H > 2 \times m_W$. At
low mass, the $b\bar{b}$ and $\tau\bar{\tau}$ decays are the dominating modes. The intermediate mass
range of $115 < m_H < 135$ GeV offers the maximal sharing of the total decay width
between the various decay channels. The decays in $c\bar{c}$ or gluon pairs are essentially
unobservable as they are overwhelmingly swamped by di-jet QCD background.
For $m_H = 125$ GeV, this takes away from observation about 11.5% of the Higgs
bosons. For this mass, the di-fermions represent about 64.0% of the decays; that is

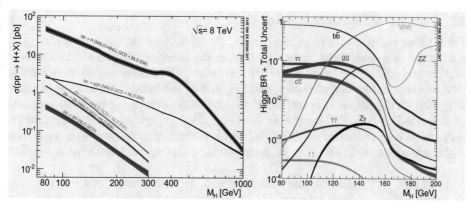

FIGURE 3. Standard model Higgs boson production cross sections at $\sqrt{s} = 8\,\text{TeV}$ (*left*). Branching ratio (BR) for the standard model Higgs boson (*right*). The plots are courtesy of Ref. [20] and reproduced here for convenience.

58 % of the Higgs bosons decaying in $b\bar{b}$ pairs, and about 6 % in $\tau\bar{\tau}$ pairs. About 24.4% branching fraction is left for the di-bosons; that is 0.228% for $\gamma\gamma$, 21.5% for WW, and 2.64% for ZZ decays. Two high mass resolution decay modes offer the best discovery potential in the intermediate masse range, the $H \to \gamma\gamma$, and the decay chain $H \to ZZ^* \to 4\ell$ (in short $H \to 4\ell$) with at least one Z boson off-mass shell and $\ell = e, \mu$. Real photons being massless, the $H \to \gamma\gamma$ decay proceeds at leading order via a fermion (mostly top quark) or boson (W) virtual loop as illustrated in Figure 4. The W loop contribution to the decay dominates. The W

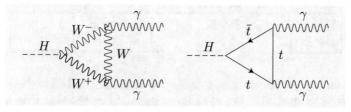

FIGURE 4. Examples of leading order Feynman diagrams contribution to the decay of the SM Higgs boson in two photons.

loop and top quark loop contributions interfere destructively such that the W+top contributions are overall about 23% smaller than the W contribution alone. While the $H \to \gamma\gamma$ decay is a rare decay mode, with its branching fraction of about 2×10^{-3} for $m_H = 125\,\text{GeV}$, the $H \to 4\ell$ decay is even rarer, with a branching fraction of about 1.2×10^{-4} for $m_H = 125\,\text{GeV}$ when considering $4\ell = 4e, 4\mu$ and $2e2\mu$ final states.

3.2. Overview of the analysis channels

For a given Higgs boson mass hypothesis, the sensitivity for the search and measurements in a given final state depends on the product of the production cross section and branching fraction to that final state, the reconstructed mass resolution, the signal selection efficiency, and the level of SM backgrounds in the relevant Higgs boson signal phase space. A list of production and decay channels explored during run I at the LHC by the ATLAS and CMS experiments, as well as an indication of the reconstructed Higgs boson mass resolution achievable in each final state, are given in Table 1. In each experiment, the pp collision events are first

TABLE 1. Production and decay channels explored during run I at the LHC. The channels labelled "$\star\star\star$" are observed by the ATLAS and/or CMS experiments and used for the determination of the Higgs boson mass and spin-parity state. Evidence is obtained for the channels labelled "$\star\star$". A sensitivity approaching the SM expectation is obtained for those labelled "$\star$". All above channels enter the ATLAS and/or CMS global combinations to constrain the Higgs boson couplings. The sensitivity is found well below SM expectation for the channels labelled with a "$\circ$". The channels labelled "- -" are out of reach.

Decay channel	$\Delta M/M$ (sub-channel)	Production Modes			
		ggH	VBF	VH	$t\bar{t}$H
H$\to \gamma\gamma$	1–2%	$\star\star\star$	$\star\star$	$\star$	$\circ$
H$\to$ ZZ*	1–2% (4ℓ)	$\star\star\star$	$\star$	$\circ$	$\circ$
H$\to$ W$^+$W$^-$	20–30% ($2\ell 2\nu$)	$\star\star\star$	$\star\star$	$\star$	$\circ$
H$\to b\bar{b}$	10–15%	- -	$\circ$	$\star\star$	$\star$
H$\to \tau^+\tau^-$	15–20%	$\star$	$\star\star$	$\star$	$\circ$
H$\to$ Z γ	1–2 %	$\star$	$\circ$	- -	- -
H$\to \mu^+\mu^-$	< 1 %	$\circ$	$\circ$	- -	- -

selected to create partitions corresponding to mutually exclusive channels. These channels are then studied in stand-alone analyses, or re-combined via a statistical method to improve the measurements of the Higgs boson properties.

The H $\to$ WW$^{(*)}$ $\to 2\ell 2\nu$, channel covers a wide mass range, but suffers from the lack of mass resolution due to the escaping neutrinos. This was the main channel used at the LHC for early searches of the Higgs boson, with a best sensitivity for a mass hypothesis around $m_{\rm H} \simeq 2 \times m_{\rm W}$. This was complemented for the search at higher mass by the H $\to$ ZZ channels (4ℓ and $2\ell 2\nu$), and at lower mass by a combination of the H $\to$ ZZ* $\to 4\ell$ and H $\to \gamma\gamma$ channels. The H $\to \gamma\gamma$ and the H $\to$ ZZ* $\to 4\ell$ channels provide a distinctive signal with a narrow peak over a smooth background. Evidence for a signal for a mass around 125 GeV in each of these channels, for both experiments, lead to the announcement of a discovery in 2012. The di-boson channels in the $\gamma\gamma$, 4ℓ, and $2\ell2\nu$ final states are the main channels that brought a significant contribution in the statistical combination contemporary to the discovery. The early searches at the LHC and the discovery of the new boson are discussed further in Section 4.

For a Higgs boson mass of about 125 GeV, all five main channels listed first in table 1, namely the di-boson channels H $\rightarrow \gamma\gamma$, H $\rightarrow$ ZZ*, and H $\rightarrow$ WW*, and the fermionic channels H $\rightarrow$ bb and H $\rightarrow \tau\tau$, can be studied at the LHC using the full run I data.

The H $\rightarrow 4\ell$ and H $\rightarrow \gamma\gamma$ channels play here again a special rôle as they provide an excellent mass resolution for the reconstructed di-photon and four-lepton final states, respectively. These channels and the Higgs boson mass measurement as well as direct constraints on the width of the resonance will be discussed in more details in Section 5.1. With a natural width of the Higgs boson expected to be in the MeV range for $m_{\mathrm{H}} \simeq 125$ GeV, and a measurement mass resolution in the GeV range, the direct measurement allows at best to conclude that the observations are consistent with a single narrow resonance. Much more stringent constraints can be obtained in an indirect manner, combining the H $\rightarrow 4\ell$ measurements at the 125 GeV resonance, corresponding to the production of a Higgs boson on mass-shell, with measurements at high mass corresponding to the exchange of a Higgs boson off mass-shell. Stringent constraints obtained on the Higgs boson intrinsic width in such a manner will be discussed in Section 5.3.

The coupling of the Higgs boson to fermions is best established directly by using the H$\rightarrow b\bar{b}$ and H$\rightarrow \tau^+\tau^-$ channels. Both channels suffer from large backgrounds and have a poor mass resolution, especially for Higgs bosons produced at low transverse momenta (P_T). For the H$\rightarrow b\bar{b}$ decay channel, a sensitivity to the signal can be enhanced by targeting the VH production mode, with V = W or Z, and with subsequent W or Z leptonic decays. For the H$\rightarrow \tau\tau$ channel, a sensitivity to the signal can be obtained by considering a combination of events with high reconstructed P_T of the Higgs boson, and events targeting the VBF production of the Higgs boson. The direct coupling to fermions and the question of flavour universality will be discussed in Section 5.4.

The measurements in all five main channels in the low mass range (110 < H < 150 GeV) can be combined to extract signal rates to be compared with SM expectations, and constraints on the Higgs boson couplings. These combined results are discussed in Section 5.5.

4. The search and the discovery

Direct and model independent searches of the Higgs boson at the LEP e^+e^- collider led to a lower bound on its mass of 114.4 GeV [21] at 95% confidence level (CL). Following the shutdown of the LEP collider in 2000, the direct search for the Higgs boson continued at Fermilab's Tevatron $p\bar{p}$ collider. The H $\rightarrow$ WW $\rightarrow 2\ell2\nu$ was the main channel used for early searches at the Tevatron, with background processes from non-resonant WW production and from top-quark production, including $t\bar{t}$ pairs and single-top-quark (mainly tW). With up to 7.1 fb^{-1} and 8.2 fb^{-1} of data from the CDF and D0 experiments respectively, the Tevatron combination [22] in 2011 excluded the range 158–173 GeV. At this time, the LHC experiments were ready to take over.

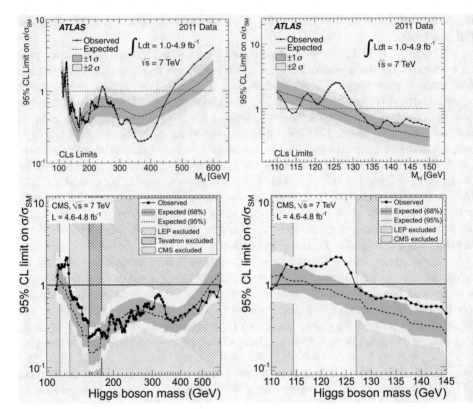

FIGURE 5. Upper limits from ATLAS (top) and CMS (bottom) using 2011 data with pp collisions at $\sqrt{s} = 7$ TeV. The 95% upper limits on the signal strength parameter $\mu = \sigma/\sigma_{SM}$ for the SM Higgs boson hypothesis is plotted as function of the Higgs boson mass.

Meanwhile, indirect constraints had been derived by exploiting the sensitivity to the Higgs boson mass of precision electroweak measurements mainly at LEP, SLC, and Tevatron colliders. A global fit [23] of the results available by the summer of 2011 suggested that the Higgs boson should have a masse below 165 GeV at 95% CL. The fit gave a best mass value of $m_H = 91^{+30}_{-23}$ GeV indicating that, in the strict context of the SM, the Higgs boson should be preferably light, if it existed.

The total production cross section at the LHC is about 20 times larger than the corresponding total cross section at the Tevatron collider for $p\bar{p}$ collisions at $\sqrt{s} = 1.96\,\text{TeV}$. With about $10\,\text{fb}^{-1}$ of data collected in the D0 and CDF experiments by the end of the Tevatron lifetime, it was expected that the ATLAS and CMS experiments at the LHC would cover previous searches and take over with less than about $1\,\text{fb}^{-1}$ of data. This occurred in 2011. As for the Tevatron, the H $\to$ WW $\to 2\ell2\nu$ was the main channel used at the LHC for early searches of the Higgs boson. By the time of the Lepton-Photon international conference in August 2011, both LHC experiments provided an exclusion at 95% CL of the Higgs boson for masses m_H around $2 \times M_\text{W}$, in a mass window extending beyond the reach of the Tevatron experiments. From the H $\to$ WW channel alone, CMS using $1.5\,\text{fb}^{-1}$ of pp collision data at $\sqrt{s} = 7\,\text{TeV}$ excluded [24] the existence of the SM Higgs boson in the range 147–194 GeV, while ATLAS using $1.7\,\text{fb}^{-1}$ of data excluded [25] the range 154–186 GeV.

By fall 2011, both LHC experiments had deployed first analyses in all main decay channels covering the full mass range. At higher Higgs boson masses, the search in the H $\to$ WW channel is complemented by the use of the H $\to$ ZZ channel. The H $\to$ WW decay has two modes (W$^+$W$^-$ and W$^-$W$^+$). Taking into account the differences in mass between the Z and W bosons, the partial width for H $\to$ ZZ is slightly less than that of one of the WW modes, i.e., less than half of H $\to$ WW. The H $\to$ ZZ nevertheless provides the best sensitivity for $M_\text{H} \gg 2 \times M_\text{Z}$ from the combination of the H $\to$ ZZ $\to 4\ell$ and H $\to$ ZZ $\to 2\ell2\nu$ channels, with $\ell = e, \mu$ and $\nu = \nu_e, \nu_\mu, \nu_\tau$. These channels were combined already at the end of the 2011 data taking campaign and this lead to the rather dramatic results shown in Figure 5. With less than $5\,\text{fb}^{-1}$ of data collected at $\sqrt{s} = 7\,\text{TeV}$ in each experiment, the full mass range for masses $m_\text{H} > 130\,\text{GeV}$ was excluded. Somehow Nature has made it as difficult as possible, possibly hiding a cherished treasure in the most inaccessible range of $114.4 < m_\text{H} < 130\,\text{GeV}$.

What followed now belongs to the history of science. Another $5\,\text{fb}^{-1}$ of data was collected at $\sqrt{s} = 8\,\text{TeV}$ until June 2012 when the experimental data was re-analysed, leading to the discovery [2, 3, 4] of a new boson around 125 GeV obtained from a combination of the di-boson channels, with leading contributions from the high resolution H $\to \gamma\gamma$ and H $\to 4\ell$ channels.

After the discovery, the landscape for the physics in relation with the Higgs boson was completely redefined. The proof of the existence of a scalar field which pervades the Universe has consequences on the history of matter, and open-up new questioning in particle physics and cosmology. Do neutrinos interact with the Higgs field? Does the Higgs boson interact with dark matter? Is there a connection between the Higgs field and the scalar field responsible for the exponential growth of the early Universe? At the LHC the research interest shifted from the search to the understanding of the exact nature of the new particle, as well as to the measurements of its properties. More than twice as much data had been collected by each of the experiments by the end of 2012. The analysis program was enriched to cover precision measurements such as its mass and width, its production cross section and quantum numbers, its couplings to other SM particles, and also searches for the rare decays such as H→ $\mu^+\mu^-$ and H→Zγ. This was enough to confirm that the new boson has properties compatible with those expected for the Higgs boson. The analysis program has been furthermore extended to cover searches for which the Higgs boson is used as a tool to probe physics beyond the SM. The results using the full run I datasets are presented below.

5. Measurements and properties

In the following, we review the results obtained by the ATLAS and CMS experiments for the new boson at 125 GeV, using all available data from the LHC run I. We first discuss the results obtained in individual bosonic and fermionic decay channels, and then the signal rates and coupling constraints obtained from a combination of all main channels.

5.1. High resolution decay channels and the Higgs boson mass

The mass of the Higgs boson is determined by combining two discovery channels with excellent mass resolution, namely H $\rightarrow$ $\gamma\gamma$ and H $\rightarrow$ ZZ* $\rightarrow$ 4ℓ. In each of these channels, the instrumental mass resolution $\Delta M/M$ is expected to be in the 1–2% range. For a SM Higgs boson resonance at a mass around 125 GeV, we expect that the intrinsic width has a negligible contribution to the measured mass resolution.

The H $\rightarrow$ $\gamma\gamma$ signal is characterized by a narrow signal mass peak over a large but smoothly falling background. The photons in background events originate from prompt non-resonant di-photon production or from jets misidentified as isolated photons. Details concerning the event selection can be found for ATLAS in Ref. [26] and for CMS in Ref. [27]. In both experiments, the analyses are split in mutually exclusive event classes to target the different production processes. The classification differs in the details between the experiments but it follows similar principles. Requiring the presence of two forward jets with high common invariant mass and a large rapidity gap favours events produced by the VBF mechanism. Event classes designed to preferentially select VH (V = W or Z) mainly require the presence of isolated electrons, muons, or missing transverse energy E_T^{miss}, or a dijet system with an invariant mass compatible with m_{W} or m_{Z}. The remaining "untagged" events correspond mainly to the Higgs boson produced via gluon fusion and represent more than 90% of the expected signal in the SM. In both experiments, the "untagged" events are further split in categories according to the kinematics of the di-photon system, and the event-by-event estimate of the di-photon mass resolution which depends on photon reconstruction in different $|\eta|$ ranges of the detectors. In total, the ATLAS and CMS analyses rely on more than 10 categories for each of the $\sqrt{s} = 7$ and 8 TeV samples. With an unfavourable signal-to-background ratio (S/B $\ll$ 1 in most categories), a key to the H $\rightarrow$ $\gamma\gamma$ analyses is the energy calibration of photons. This is obtained by using the Z→ ee candle and extrapolating to the relevant p_T range of photons, taking into account the effects from the different

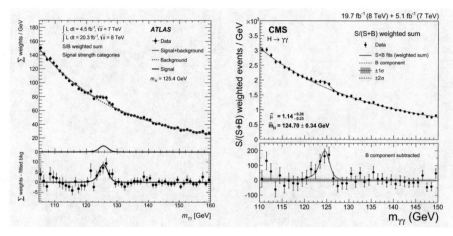

FIGURE 6. Distribution of the di-photon invariant mass measured in the H → γγ analyses for run I data at 7 and 8 TeV. Combination of the event classes showing weighted data points with errors, and the result of the simultaneous fit to all categories from (*left*) ATLAS and (*right*) CMS experiments. In each case, the fitted signal plus background is shown along with the background-only component of this fit together, and the background subtracted weighted mass spectrum is shown in the bottom.

behaviour of photon-induced and electron-induced electromagnetic showers (e.g., shift of the longitudinal profile) in the detector. Overall, the analyses have an acceptance×efficiency of about 50% and the event categorisation is expected to improve the sensitivity by about a factor two with respect to a fully inclusive analysis. The di-photon invariant mass distribution measured by the experiments is shown in Figure 6. A clear Higgs boson signal resonance is observed around 125 GeV. ATLAS observes [26] a signal with a local significance of 5.2σ, for a SM Higgs boson expectation of 4.6σ, at the mass obtained by the combining of the 4ℓ and 2γ channels [28]. CMS observes [27] a signal with a local significance of 5.7σ, for a SM Higgs boson expectation of 5.2σ, at the mass measured in the $\gamma\gamma$ channel in stand-alone.

The H → ZZ* → 4ℓ signal is characterized by a narrow four-lepton ($4e, 2e2\mu$ or 4μ) mass peak over a small continuum background. Details concerning the event selection in this channel can be found for ATLAS in Ref. [29] and for CMS in Ref. [30]. The ATLAS and CMS analyses differ in the details but follow similar principles. The signal candidates are divided into mutually exclusive quadruplet categories, $4e$, $2e2\mu$ and 4μ, to better exploit the different mass resolutions and different background rates arising from jets misidentified as leptons. Four well-identified and isolated leptons are required to originate from the primary interaction vertex to suppress the Z+ jet and $t\bar{t}$ instrumental backgrounds. With

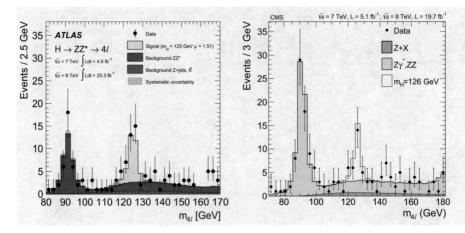

FIGURE 7. Distribution of the four-lepton invariant mass measured in the $H \rightarrow Z^{(*)}Z^* \rightarrow 4\ell$ analyses for run I data at 7 and 8 TeV, from (*left*) ATLAS and (*right*) CMS experiments. The plots show the sum of the 4e, 2e2μ and 4μ channels, with points with error bars representing the data, and shaded histograms representing the backgrounds. Superimposed in each case is an histogram for the Higgs boson signal expectation. This signal expectation is shown for a mass $m_H = 125$ GeV and a signal strength $\mu = \sigma_{\text{obs.}}/\sigma_{\text{SM}} = 1.51$ in the case of ATLAS, and for $m_H = 126$ GeV and the standard model expectation ($\mu = 1.00$) in the case of CMS.

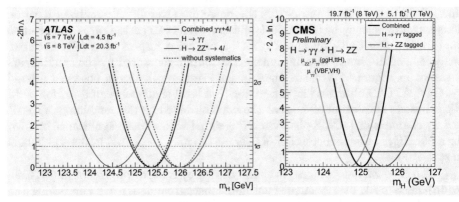

FIGURE 8. Scan of the likelihood test statistic versus the Higgs boson mass m_H for the $H \rightarrow \gamma\gamma$ and the $H \rightarrow 4\ell$ channels, and their combination, for (*left*) ATLAS, and (*right*) CMS.

a very favourable expected signal-to-background ratio (S/B $\gg$ 1), a key to the H $\to$ 4ℓ analyses is to preserve the overall efficiency while imposing lepton identification and isolation criteria sufficient to suppress the instrumental background well below the indistinguishable background from the non-resonant ZZ continuum. The fourth lepton (i.e., with lowest p_T) has its p_T peaking well below 10 GeV for $m_H = 125$ GeV. A high lepton reconstruction efficiency is required down to the lowest p_T consistent with the rejection of instrumental background; in practice the lowest threshold is in the range 5 to 7 GeV. The electron reconstruction makes use of rather sophisticated algorithms which combine the reconstructed track in the silicon tracker (using a gaussian sum filter technique dedicated to electrons) with clusters in the electromagnetic calorimeter, a categorization of electrons, etc. The energy scale is controlled using the Z$\to \ell\ell$ candle complemented by the validation at low p_T from J/ψ and Υ(nS). The signal candidates should contain two pairs of same flavour and opposite charge leptons ($\ell^+\ell^-$ and $\ell'^+\ell'^-$). For $m_H = 125$ GeV, the decay H $\to$ Z$^{(*)}$Z* involves at least one Z boson off mass-shell (i.e., ZZ*), and, for about 20% of the cross section, two Z boson off mass-shell (i.e., Z*Z*). The analysis thus accepts a leading Z boson (Z_1) reconstructed with masses down to 40 or 50 GeV, and a subleading one (Z_2) with masses down to 12 GeV. Overall, the analyses have an acceptance$\times$efficiency of about 20 to 40% depending on the quadruplet category. Even more sophisticated statistical analysis techniques are used beyond the baseline selection of signal candidates. In CMS, kinematic discriminants are constructed using the masses of the two di-lepton pairs and five angles, which uniquely define a four-lepton configuration in their centre-of-mass frame. These make use of leading order matrix elements for the signal and background hypothesis and are used to further separate signal and background. In ATLAS the analysis sensitivity is improved by employing a multivariate discriminant to distinguish between the Higgs boson signal from the ZZ^* background and the combination of this discriminant with the reconstructed invariant mass of the 4ℓ system is used to provide the final measurement on the Higgs boson mass in this channel.

TABLE 2. Signal strengths and mass measurements from the high resolution di-boson channels at the LHC.

Expt.	Decay Channel	Signal Strength $\mu = \sigma_{\text{meas.}}/\sigma_{SM}$	Measured Mass (GeV) mass $\pm$ statistics $\pm$ systematics	Reference
ATLAS	H$\to \gamma\gamma$	$1.29^{+0.30}_{-0.30}$	125.98 ± 0.42(stat) ± 0.28(syst)	[28]
	H$\to$ZZ$^*\to 4\ell$	$1.66^{+0.45}_{-0.38}$	124.51 ± 0.52(stat) ± 0.06(syst)	[28]
	Combined	–	**125.36 $\pm$ 0.41**	[28]
CMS	H$\to \gamma\gamma$	$1.14^{+0.26}_{-0.23}$	124.7 ± 0.31(stat) ± 0.15(syst)	[27]
	H$\to$ZZ$^*\to 4\ell$	$0.93^{+0.29}_{-0.25}$	125.6 ± 0.4(stat) ± 0.2(syst)	[30]
	Combined	–	**125.02 $\pm$ 0.30**	[31]

The 4ℓ invariant mass distribution measured by the experiments is shown in Figure 7. One observes a very clear Higgs boson resonance over a smooth background. The signal is observed with very high significance in both experiments. CMS observes [30] a signal with a local significance of 6.8σ, for a SM Higgs boson expectation of 6.7σ, at the mass measured in the 4ℓ channel in stand-alone. ATLAS observes [29] a signal with a local significance exceeding 8σ, for a SM Higgs boson expectation of 6.2σ, at the mass obtained by combining the 4ℓ and 2γ channels.

The measurements of the Higgs boson mass in the $\gamma\gamma$ and 4ℓ channels and for their combination are listed in Table 2, and shown in Figure 8. These final run I measurements profit from the most accurate knowledge of the detector performance achieved so far, using the full datasets from proton proton collisions at the LHC in 2011 and 2012. The mass measured in the $\gamma\gamma$ channel is obtained in both experiments via a simultaneous fit of all event categories. The mass measured in the 4ℓ channel is obtained by ATLAS using a "2D" fit combining the reconstructed mass and a BDT discriminant trained on signal and ZZ* background events from Monte Carlo simulation. The mass measured in the 4ℓ channel by CMS uses a "3D" fit combining the reconstructed mass, a kinematic discriminant based on matrix elements tuned to distinguish signal from ZZ* background, and the uncertainty in the four-lepton mass estimated from detector information on a per-event basis. This is found relevant for CMS because this uncertainty varies considerably over the small number of selected signal events. In both experiments, the precision on the mass measurements from both the $\gamma\gamma$ and 4ℓ channels suffers mainly from the limited statistics. The new data taking campaign at the LHC starting in 2015 will be important to decrease the uncertainty in this measurement. A final mass value is obtained by combining the $\gamma\gamma$ and 4ℓ results. ATLAS obtains [28] a mass of $m_\mathrm{H} = 125.36 \pm 0.37(\mathrm{stat}) \pm 0.18(\mathrm{syst})\,\mathrm{GeV}$ (i.e., 125.36 ± 0.41). CMS obtains [31] a mass of $m_\mathrm{H} = 125.02^{+0.26}_{-0.27}(\mathrm{stat})\,^{+0.14}_{-0.15}(\mathrm{syst})\,\mathrm{GeV}$ (i.e., 125.02 ± 0.30). The results are found to be consistent between channels within each experiment, and remarkably similar between the experiments for the final mass values. One notices the few per-mil level of accuracy achieved in this measurement.

5.2. The Higgs boson intrinsic width

The intrinsic width (Γ_H) of the Higgs boson in the SM is $\Gamma_\mathrm{H} \simeq 4.2\,\mathrm{MeV}$ for $m_\mathrm{H} = 125\,\mathrm{GeV}$, corresponding to a lifetime $\tau^0_\mathrm{H} = \hbar/\Gamma_\mathrm{H} \simeq 2 \times 10^{-22} s$. This Γ_H is too small for a direct observation at the peak where the measured width is completely dominated by detector resolution, while at the same time too large to allow for the observation of displaced vertices via its lifetime. At best, the experiment can verify that the lineshape at the resonance is consistent with a single narrow resonance. This has been explicitly done by both the ATLAS [26, 32] and the CMS experiments [27, 30]. ATLAS sets direct limits at 95% CL of $\Gamma_\mathrm{H} < 5\,\mathrm{GeV}$ from the H $\to \gamma\gamma$ channel, and $\Gamma_\mathrm{H} < 2.6\,\mathrm{GeV}$ from the H $\to 4\ell$ channel. CMS sets direct limits at 95% CL of $\Gamma_\mathrm{H} < 2.4\,\mathrm{GeV}$ from the H $\to \gamma\gamma$ channel, and $\Gamma_\mathrm{H} < 3.4\,\mathrm{GeV}$ from the H $\to 4\ell$ channel. A sensitivity to a range of intrinsic width values of the order of Γ_H is nevertheless possible by profiting from the fact that the narrow width approximation fails for the production of a Higgs boson via gluon fusion (ggH). The off-shell production cross section is sizeable and this has been exploited by the experiments in the ZZ and WW channel.

In the ZZ channel for instance, sizeable off-shell production of the Higgs boson arises from an enhancement in the decay amplitude in the vicinity of the Z-boson pair production threshold, and at higher masses from the top-quark pair production threshold. There is in addition at large mass a sizeable destructive interference with the production of a Z-boson pair from the continuum (i.e., with Z bosons coupling to quarks in a box diagram). Overall the ratio of the off-shell (above $2 \times m_Z$) to the on-shell cross section is of the order of 8%. This sizeable contribution of the off-shell Higgs boson is not as such surprising. The Higgs boson is essential for the unitarity of the theory and it must be there to play its role in canceling the bad high energy behaviour of the continuum diagrams. The on-shell and off-shell cross section can be approximated as:

$$\sigma^{\text{on-shell}}_{gg \to H \to ZZ^*} \approx \frac{g^2_{gg\text{H}} g^2_{HZZ}}{m_\text{H} \Gamma_\text{H}} \quad \text{and} \quad \sigma^{\text{off-shell}}_{gg \to H^* \to ZZ^*} \approx \frac{g^2_{gg\text{H}} g^2_{HZZ}}{2m_Z} \ .$$

Thus, a measurement of the relative off-shell to on-shell signal production in the ZZ channel provides direct information on Γ_H. Using this idea [33, 34, 35, 36], the CMS experiment has obtained [37] a constraint on the total width of $\Gamma_\text{H} < 22$ MeV (i.e., 5.4 times the expected value in the SM) at 95% CL. In a similar analysis ATLAS has obtained [38] a constraint at 24 MeV (5.7 times the expected value in the SM) at 95% CL.

5.3. The Higgs boson spin-parity

Extensive tests of the spin-parity state of the new boson at the LHC have been performed in the di-boson decay channels by both the ATLAS and CMS experiments. The tests in the H $\to \gamma\gamma$ channel exploit the production dependent scattering angle of the di-photon pair. The experiments make use of the absolute value of the cosine of the polar angle of the di-photons in the Collins–Sopper [39] rest frame, i.e., the angle in the di-photon rest frame between the collinear photons and the line that bisects the acute angle between the colliding protons. The analysis takes into account the differences between the angular distribution expected for the production of the resonance via $q\bar{q}$ annihilation and for the production via a mixture of gg fusion and $q\bar{q}$. The tests in the H $\to$ ZZ$^* \to 4\ell$ channels exploit the masses and angles reconstructed from the four leptons. The kinematic properties of the SM Higgs boson or any non-SM exotic boson decay to the four-lepton final state has been extensively studied in the litterature (see, e.g., Ref. [27] for a complete set of references), and can be described by the reconstructed masses and five production and decay angles. The tests in the H $\to$ WW$^* \to 2\ell2\nu$ channels are production dependent and exploit a combination of observables such as the di-lepton invariant mass $m_{\ell\ell}$, the azimuthal separation between the two leptons $\Delta\phi_{\ell\ell}$, the di-lepton transverse momentum $p_{T\ell\ell}$, and the reconstructed transverse mass m_T. ATLAS combines the sensitive observables in a BDT. CMS uses the "2D" distribution in the plane $m_{\ell\ell}$ vs. m_T, where m_T expressed in terms of the $p_{T\ell\ell}$, $\Delta\phi_{\ell\ell}$, and the missing transverse momentum in the event. In all cases, the hypothesis of a SM Higgs boson in a pure spin-parity state $J^P = 0^+$ is compared with alternative J^P hypotheses. The spin 1 is excluded in principle by the Landau–Yang theorem and the observation of the H $\rightarrow \gamma\gamma$ channel. The observation of the new boson in this channel implies that the resonance must be a boson with spin 0 or 2. Exotic spin-1 hypotheses have been nevertheless tested in the ZZ* and WW* channels. Binned likelihood fits are used to test the data for compatibility with the presence of a particle with given spin-parity J^P.

In all cases, the data is found compatible with the $J^P = 0^+$ quantum numbers of the Higgs boson, whereas alternative hypotheses are excluded with high confidence levels. The CMS results from individual di-boson channels are described in Refs. [40, 30, 27, 41]. CMS excludes $J^P = 0^-$ and $1^+, 1^-$ hypotheses at 99% CL or higher, and the spin-2 hypotheses at 95% CL or higher. The ATLAS spin-parity results for di-boson channels are described in Ref. [42]. ATLAS excludes the $J^P = 0^-, 1^+, 1^-, 2^+$ hypotheses at 97.8% CL or above. For both experiments, the exclusions are found to hold independently of the assumptions on the coupling strengths to the SM particles, and, for spin-1 and spin-2 hypothesis, of the relative fraction of gluon fusion and quark-antiquark production. Further constraints on pure and mixed spin-parity states under various assumptions have been recently established by CMS combining all di-boson channels [43]. Overall, the data thus provide strong evidence for the spin-0 nature of the Higgs boson, with positive parity being strongly preferred. The CP-even 0^+ hypothesis is found to be favoured over any other pure spin-parity state hypothesis at a level of more than 3 standard deviations.

5.4. The fermionic decay modes and non-universality

The H$\rightarrow b\bar{b}$ decay channel is studied in the VH production mode with V = W or Z, and with V undergoing leptonic decays [44, 45]. Final states with 2 b jets from the H decay, and with zero, one, or two charged leptons (electrons or muons) from the V decays are considered by the experiments, targeting Z$\rightarrow \nu\nu$, W$\rightarrow e\nu, \mu\nu$, and Z$\rightarrow ee, \mu\mu$ respectively. The channel W$\rightarrow \tau\nu$ is also considered in CMS in the case where the τ decay involves one charged hadron, i.e., the so-called "single-prong" decays. The key elements of the analysis are to obtain a high efficiency in tagging the b-jets, a low rate of misidentified jets as b-jets, and an estimation of the backgrounds from the data. Requirements on the missing transverse energy and/or on the azimuthal opening angle between the missing transverse momentum and the direction of the b jets (or of the leptons) are imposed. To further improve on the sensitivity, the analysis for each final state is further divided in categories according to the p_T boost of the H or the V bosons. The H and the V bosons recoil against each other and a substantial reduction of the background can be achieved in high p_T boost kinematic regions [46]. For the statistical analysis of the selected events, ATLAS employs a binned likelihood constructed as the product of distributions for the invariant mass $m_{b\bar{b}}$ in 26 signal regions, while CMS employs a combination of 14 boosted-decision tree (BDT) discriminants. While signal over background (S/B) ratios in the range of 0.1% to 1.0% are expected when integrating around the signal peak at $m_{b\bar{b}} \simeq 125\,\text{GeV}$, this improves up to about 10% for events with highest BDT scores. The $t\bar{t}$ production is among the main backgrounds in all event categories. It dominates the event yield in the signal region for WH production after the full event selection. The V+$b\bar{b}$ production is the dominating background for ZH production.

The H$\rightarrow \tau^+\tau^-$ decay channel is studied in the ggH, VBF, and VH production modes [47, 48], with $\tau_\ell\tau_\ell, \tau_\ell\tau_h$ and $\tau_h\tau_h$ in the final state, where $\tau_\ell = \tau_e$ or τ_μ designates tau leptons decaying leptonically, and τ_h designates tau leptons decaying semi-leptonically (with one or more charged hadrons in the final state). To enhance the sensitivity in the ggH or VBH production modes, the events are classified in categories according to the number of additional jets and to kinematic quantities that exhibit differences for the signal and background events. Categories with large p_T (boosted) reconstructed Higgs boson enhance the sensitivity to ggH production. Categories with 2 high p_T jets separated by a large rapidity gap target VBF production. CMS also considers VH production, requiring one or two additional leptons (electrons or muons) compatible with a leptonic decay of the W or Z boson. For the statistical analysis of the selected events, ATLAS uses a combination of BDTs, built in the various $\tau_\ell\tau_\ell, \tau_\ell\tau_h$ and $\tau_h\tau_h$ channels from a set of discriminating variables, to combine the VBF and boosted exclusive categories. CMS employs a likelihood product with the signal extracted in the different channels from the distribution of the invariant mass of the tau lepton pair, except in the WH and in the ee and $e\mu$ channels where kinematics discriminants are used. Signal over background (S/B) ratios or the order of 10% are achieved in the three bins with highest BDT score, and reaches S/B$\simeq$1 for the VBF bin with highest BDT score of ATLAS (as for the "tight VBF category" of CMS) where 10–20 signal events are expected.

In the H$\rightarrow b\bar{b}$ channel, ATLAS observes [44] a 1.4σ excess with respect to the background only hypothesis, for an expectation of 2.6σ for the SM Higgs boson. CMS observes [45] an excess of 2.1σ compared to an expectation for the SM Higgs boson of 2.1σ. The statistics in the VH production mode is too small at the LHC to establish at this stage a direct evidence for H$\rightarrow bb$. The most significant evidence so far for H$\rightarrow b\bar{b}$ comes from the CDF and D0 experiments at the Tevatron. Combining their analyses in the VH production modes, the Tevatron experiments [49, 50] find an excess of signal candidates with a significance of 2.8σ at the LHC mass $m_H = 125\,\text{GeV}$, and a maximum local significance of 3.3σ at 135 GeV.

In the H$\rightarrow \tau^+\tau^-$ channel, both LHC experiments find clear evidence for a Higgs boson signal [47, 48], thus establishing, beyond the knowledge available at the time of the discovery, the first evidence that the Higgs boson couples to leptons. CMS finds [45] a 3.4σ excess with respect to the background only hypothesis, for an expectation of 3.6σ for the SM Higgs boson. The observations in the various event categories used for the analysis are illustrated in Figure 9. A combination of

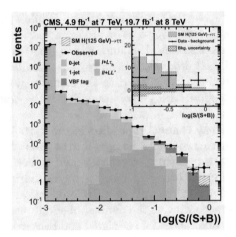

FIGURE 9. Combined observed and predicted distributions for the H→ $\tau^+\tau^-$ observations by the CMS experiment [48]. Similar results are obtained by the ATLAS experiment [47]. The results are presented grouped in bins of the $\log(S/S+B)$ for the final discriminators used for the various event categories, with $(S/S+B)$ denoting the ratio of the predicted signal (S) and signal-plus-background $(S+B)$ event yields in each bin.

the $b\bar{b}$ and $\tau^+\tau^-$ decay channels [51] yields an evidence for the coupling to these fermions at 3.8σ (4.4σ expected). In the H→ $\tau^+\tau^-$ channel, ATLAS finds [44] a 4.1σ excess for an expectation of 3.2σ for the SM Higgs boson. The evidence for the H$\tau\tau$ coupling combined with the null evidence so far for the H$\mu\mu$ coupling [52, 53] implies that the new boson has non-universal family couplings. The scalar sector could play an important role in the origin of fermion families.

5.5. Combined measurements of signal rates and couplings

A coherent statistical analysis of the full set of analysis channels allows to slightly improve the measurements of the signal rates for individual production and decay modes of the Higgs boson, as well as to establish a coherent set of constraints on the Higgs boson couplings to different particle species.

For the measurement of signal rates, the inputs to the combined analyses are in principle the experimental results obtained in individual, i.e., "stand-alone" and mutually exclusive analyses discussed in previous sections of this paper. In practice, the ATLAS and CMS combinations make use of more or different information, and uses the information from individual channels in a different manner. At this stage, this is partly because only preliminary combination results are available. ATLAS first published in summer 2013 a combination of di-boson channels using all available run I data [29], but the results in individual decay channels have been since then superseded in some cases. A new combination of the five main decays channels and preliminary results in individual channels is now available [54], but this one does not yet include for instance the latest H → γγ from ATLAS [26]. CMS first published final sets of results using all available run I data in each of the main five decays channels, and presented a preliminary combination in summer 2014 [31] which incorporate additional information analysis targeting specific production modes such as $t\bar{t}$H [55]. Also, the combination profits from new theory information in some areas. For instance, since the publication of the stand-alone analysis, the search for VH production with H → b$\bar{\text{b}}$ decays has been improved for CMS [31] by the use of recent NLO calculations for the gluon fusion loop contribution to ZH production. The organization of the information also differ for some individual decay channels. For CMS the input to the combination is organized in terms of decay "tags". For instance the H → ττ "tag" includes some signal contamination from H → WW*, etc. The signal strengths for such decay "tags" which serve as inputs to the combination in CMS cannot be interpreted literally as compatibility tests for pure production mechanisms or decay modes, in contrast to the results from the stand-alone analysis.

The γγ, ZZ, and WW di-boson channels were the main contributors to the original discovery and have been exploited for the determination of the Higgs boson mass, intrinsic width, and spin-parity state. For the combination, both the ATLAS and CMS experiment assume a single CP even scalar state (0^+) resonance with a mass obtained by the combination of the H → γγ and H → ZZ* → 4ℓ ($\ell = e, \mu$) channels, as established from the di-boson decay channels discussed in previous sections. While the mass of the Higgs boson is a free parameter in the SM, the number of Higgs bosons events decaying in each channel is quite accurately predicted by theory; thus measuring the ratio μ between the number of observed events over the number of predicted events (signal strength) we have an easy way to test the consistency with the SM ($\mu = 1$ means that what we observe is the SM Higgs boson). To obtain specific constraints on the Higgs boson couplings, a simultaneous analysis of all production and decay channels is necessary to account in a consistent manner for all statistical uncertainties, systematic certainties, and their correlations. Furthermore, the production×decay for the Higgs boson at the LHC is always sensitive to a combination, linear at LO, of two couplings. Thus some model assumptions are required to disentangle the effects of each coupling. This is done following the prescription of the LHC Cross Section Working group. A narrow width approximation such that $\sigma \times \beta_i = \sigma_i \times \Gamma_i / \Gamma_H$ is considered and SM "kappa" modifiers are introduced for the production, $\kappa_i^2 = \sigma_i / \sigma_i^{SM}$, and decay $\kappa_j^2 = \Gamma_j / \Gamma_i^{SM}$, with $\kappa_H = (\sum \kappa_j^2 \Gamma_j^{SM}) / \Gamma_H^{SM}$. Various benchmark scenarios are then studied [54, 31].

The signal strength μ measured in various decay channels by ATLAS [54] and CMS [31] experiments is shown in Figure 10. In both experiments, all signal strengths measured are consistent with the expectation for the Higgs boson in the SM within one to two standard deviations. The best fit signal strengths μ for di-bosons measured in ATLAS are seen in Figure 10 (left) to be slightly above

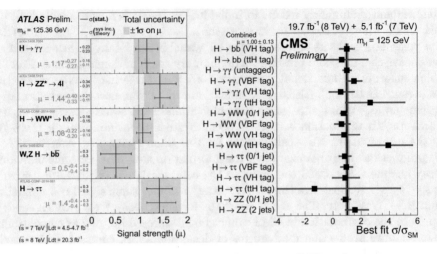

FIGURE 10. The signal strength μ at the measured Higgs boson masses by the (*left*) ATLAS and (*right*) CMS experiments. For ATLAS the best-fit values are shown by the solid vertical lines with ± 1 standard deviation uncertainties indicated by green shaded bands, and the contributions from statistical uncertainty (top), the total (experimental and theoretical) systematic uncertainty (bottom) indicated within the bands. For CMS, the best fit value for the combination is shown as a solid vertical line and the overall uncertainty as a vertical band; the points are the results from sub-combinations by predominant decay mode or production mode tag. The uncertainties include both statistical and systematic uncertainties.

expectations, with 1.17 ± 0.27 (H $\rightarrow \gamma\gamma$), $1.44^{+0.40}_{-0.33}$ (H $\rightarrow$ ZZ*), $1.08^{+0.22}_{-0.20}$ (H $\rightarrow$ WW*). A signal strength of 1.4 ± 0.4 is obtained for H $\rightarrow \tau\tau$. A value of 0.5 ± 0.4 is obtained for H $\rightarrow$ b$\bar{\text{b}}$. Combining all all five main decay channels, using their previous H $\rightarrow \gamma\gamma$ result [29], ATLAS finds $\mu = 1.30 \pm 0.12(stat)^{+0.14}_{-0.11}(syst)$. The Figure 10 (right) show the signal strengths in the various "tags" from CMS. The signal strengths μ combining the various "tags" obtained for each of the 5 main decay channel tags are 1.13 ± 0.24 (H $\rightarrow \gamma\gamma$), 1.00 ± 0.29 (H $\rightarrow$ ZZ*), 0.83 ± 0.21 (H $\rightarrow$ WW*), 0.91 ± 0.27 (H $\rightarrow \tau\tau$), and 0.93 ± 0.49 (H $\rightarrow$ b$\bar{\text{b}}$). Combining all five main decay channels, CMS finds $\mu = 1.00 \pm 0.13$. The top quark is involved in virtual loops for the ggH production, the main production channel at the LHC, as well as in virtual loops for H $\rightarrow \gamma\gamma$ decay where it interferes with loops involving the W boson. Indirect evidence for the Higgs boson coupling to the top quark is thus obtained. The other heavy fermions of the third generation, the b quark and the τ lepton, are involved in the dominating Higgs boson decay modes. Evidence has been found for the H$\rightarrow \tau\tau$ decay as was discussed in previous sections. The bottom quark is involved mainly in the decay $H \to b\bar{b}$ where only a small excess of events has been observed so far.

The data from different decay channels can be associated to different production "tags" as was shown for example in Figure 10 (right). Each production mechanism can be in turn associated to either fermion couplings (ggH, $t\bar{t}$H) or to vector-boson couplings (VBF, VH). From the combined fit, the signal strength for the VH and VBF production can be assessed. An evidence is obtained for the observation of VBF production with a significance of 4.1σ for ATLAS [29, 54], and 3.7σ for CMS [31]. For the VH production, CMS observes a significance of 2.7σ for an expectation of 2.9σ [31].

The ATLAS and CMS constraints on the Higgs boson coupling to fermions $(\kappa_F = \kappa_\ell = \kappa_q)$ and electroweak bosons $(\kappa_V = \kappa_W = \kappa_Z)$ are shown in Figure 11. The data are compatible with the expectation for the SM Higgs boson: the SM

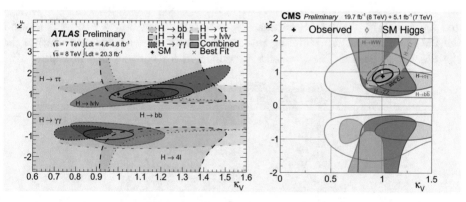

FIGURE 11. The 68% contours for individual decay channels (bounded colored regions) and for the overall combination (thick curves) in the correlation plane (κ_V, κ_F), the coupling scale factors for bosons (κ_V) and fermions (κ_F), from (left) ATLAS and (right) CMS. The standard model expectation is indicated at $(\kappa_V, \kappa_F) = (1, 1)$. The likelihood scans are shown in the two quadrants, assuming either like signs $(+, +)$ or unlike signs $(+, -)$.

point of $(\kappa_V, \kappa_F) = (1, 1)$ is within the 95%CL contour for ATLAS, and within the 68%CL contour for CMS. The fits in Figure 11 are shown allowing for opposite signs of the κ_V and κ_F. The sensitivity to this relative sign comes from the negative interference between the loop contributions involving either W bosons or top quarks in the $H \to \gamma\gamma$ decay. In other words, at LO $\Gamma_{\gamma\gamma}$ involves a product $\kappa_W \kappa_t$ while all other partial decay widths scale as κ_V^2 or κ_F^2. The $(\kappa_V, \kappa_F) = (1, -1)$ is found to be disfavoured at the $\sim 2\sigma$ level by ATLAS results, and at the $\sim 3\sigma$ level by CMS results.

The combined data has been used to further test the compatibility of the observations with the SM Higgs boson couplings, by fitting to a subset of coupling modifiers. In the SM, the custodial symmetry fixes the relative couplings $\lambda_{WZ} = \kappa_W/\kappa_Z$ of the Higgs boson to W and Z bosons to $\lambda_{WZ} = 1.0$. ATLAS obtains a best fit for this ratio of $0.94^{+0.14}_{-0.29}$ while CMS obtains $0.91^{+0.14}_{-0.12}$. From a fit for the couplings to bosons κ_V and fermions as free parameters, ATLAS obtains a best fit of $(\kappa_V, \kappa_F) = (1.15^{+0.08}_{-0.08}, 0.99^{+0.17}_{-0.15})$ while CMS obtains $(\kappa_V, \kappa_F) = (1.01^{+0.07}_{-0.07}, 0.89^{+0.14}_{-0.13})$. For the effective couplings to gluons κ_g and photons κ_γ as free parameters, ATLAS obtains best fit values of $\kappa_g = 1.08^{+0.15}_{-0.13}$ and $\kappa_\gamma = 1.19^{+0.15}_{-0.12}$, while CMS obtain $\kappa_g = 0.89^{+0.10}_{-0.10}$ and $\kappa_\gamma = 1.15^{+0.13}_{-0.13}$. Fits allowing for a different ratio of the couplings to down-type and up-type fermions $(\lambda_{du} = \kappa_d/\kappa_u)$ or, separately, for a different ratio of the couplings to leptons and quarks $(\lambda_{\ell q} = \kappa_\ell/\kappa_q)$ have been performed. These fits are motivated by theories beyond the standard model (BSM) where the couplings to different type of fermions can be modified, such in supersymmetric models. CMS obtains $\lambda_{du} = 1.01^{+0.20}_{-0.19}$ and $\lambda_{\ell q} = 1.02^{+0.22}_{-0.21}$. ATLAS obtains $\lambda_{du} = 0.95^{+0.20}_{-0.18}$ and $\lambda_{\ell q} = 1.22^{+0.28}_{-0.24}$ around the SM-like minima. All coupling results are consistent with the expectation for the SM Higgs boson. These results are collected for convenience in Figure 12.

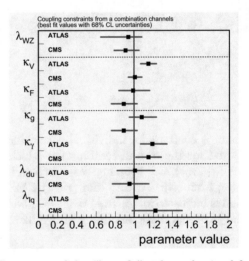

FIGURE 12. Summary of the "best fit" values obtained for different fits to the full set of Higgs boson analysis channels.

Figure 13 illustrates the results from the best fit values for the couplings for different particle species. The couplings are found to scale with mass as expected for a Higgs boson of the BEH mechanism. Finally, constraints can be obtained on possible BSM contributions by allowing for a non-vanishing partial width into invisible or undetected particles. Upper limits at 95% CL for the branching ratio into such BSM particles of 41% and 58% are obtained by ATLAS and CMS respectively.

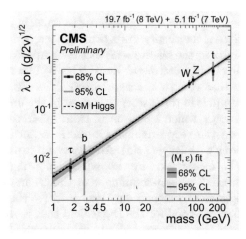

FIGURE 13. Summary of the "best fit" values obtained for different particle species, and expressed as function of the known particle mass. The results are shown for illustration here from CMS [31]. Similar results are obtained by ATLAS [54]. For the fermions, the values of the fitted Yukawa couplings Hff are shown. For the vector bosons, the square-root of the coupling for the HVV vertex, divided by two times the vacuum expectation value of the Higgs boson field are shown.

6. Conclusions and the aftermath

The boson discovered in 2012 at the LHC by the ATLAS and CMS experiments has properties consistent, within uncertainties, with the Higgs boson of the standard model. The analysis of the full sets of data collected during run I at the LHC at 7 and 8 TeV has allowed for considerable progress in the characterization of this H boson. Having determined the Higgs boson mass m_H with a relative precision $\Delta m/m$ at the few per-mil level, all the production and decay properties of a H boson are predicted by the theory and can be compared with data. The custodial symmetry is verified to $\sim 15\%$. The relative couplings of the H boson with d-like and u-like quarks of the third generation is verified at the $\sim 30\%$ level. The couplings to fermions of the third generation is verified at the ~ 15–20% level. Overall, the couplings to boson and fermions are verified to $\sim 15\%$ and consistent with the SM expectation at the $\sim 1\sigma$ level, and, thus, scale as expected as a function of the fermion and vector boson masses. The existence of a boson with non-universal family couplings is established via the evidence for H$\to \tau\tau$ and the null evidence for H$\to \mu\mu$. More data will be needed to further disentangle the various production and decay modes, and provide more stringent constraints on the couplings.

The existence of a scalar field and the spontaneous electroweak symmetry breaking mechanism, provide an explanation for the origin of the Z and W and ordinary fermion masses and solves, or postpones to much higher energy, the problem of the unitarization of the theory. As a consequence, stringent constraints are established for global fits in the electroweak sector by injecting the H boson mass in the global fit of precision electroweak data: The W boson is predicted with a precision better than that of direct measurements. The discovery marks the triumph of the weak couplings in the history of matter in the universe; a culmination of a reductionism strategy which has evolved from questions of the structure of matter to questions on the very origin of interactions (local gauge symmetries) and matter (interactions with Higgs field). An important aspect of the theory nevertheless remains to be constrained by the experiments, namely the coupling of the Higgs field with itself. This self-coupling is at the origin of the so-called "condensation" of the Higgs field which is expected to drive the EWSB mechanism. The shape of the scalar potential for the Higgs field that is responsible for EWSB depends on m_H and on the trilinear and quadrilinear self-couplings. In the SM, these are presumed to be fundamentally related. The trilinear coupling for the physical Higgs boson which enters for instance in di-H production is given in the SM by $\lambda_{HHH} = 3m_H^2/v$, where $v = (\sqrt{2}G_F)^{-1/2} \approx 246\,\text{GeV}$ is the mean vacuum expectation value for the Higgs field. The observation of the di-H production and extraction of constraints on λ_{HHH} is on the menu for the high luminosity runs at the LHC during the next five years, and a measurement is reachable at very high luminosity with a future upgraded LHC collider.

Besides the mass, spin-parity, and couplings of the Higgs boson, there still remain the questions of the origin and stabilization of its mass at the weak scale. This question of a "natural" stabilization of the Higgs boson mass had been a central incentive for the developments of theories beyond the standard model (BSM) for many decades. In so-called "Technicolor" theories, one assumes that the SM is only an effective theory which breaks up at the TeV scale where a new strong interaction sets in. In so-called "extra dimension" theories, the validity of the SM is assumed to be limited at the TeV scale where strong effects of quantum gravity propagating in all dimensions would set in. Supersymmetric theories offer in principle a more satisfactory solution in the scalar sector. The self-coupling can possibly be expressed in a combination of gauge couplings in such theories such that the scalar sector is strongly constrained, e.g., with a predicted mass for the lightest, possibly SM-like, neutral Higgs boson. The stabilization of the Higgs scalar boson is obtained, despite the introduction of the new scale for the breaking of the supersymmetry, by exact cancellations of the contributions of the new supersymmetric particles, the partners of ordinary fermions and bosons.

In the years to come, the LHC collider with operate at higher instantaneous luminosities, and higher pp centre-of-mass energies. The aim is to reach $300\,\text{fb}^{-1}$ of integrated luminosity at $\sqrt{s} = 13\,\text{TeV}$ in a first phase, and then $3000\,\text{fb}^{-1}$ at $\sqrt{s} = 14\,\text{TeV}$ in a second phase. These forthcoming data taking periods could allow for the observation of deviations from expectation or for the direct discovery of extra structure in the scalar sector, beyond the minimal sector of the standard model.

Acknowledgment

I wish to thank my colleagues from the ATLAS and CMS collaborations for their support in preparing this review. Thanks to Olivier Davignon and Roberto Salerno from the CMS group at the Laboratoire Leprince-Ringuet for comments and suggestions. This review borrows and extends from a previous review that I prepared on behalf of ATLAS and CMS experiments for the "25th Rencontres de Blois" in 2013, as well as for a more recent review in preparation with Rosy Nicolaidou from the IRFU, CEA-Saclay, for the "Comptes Rendues de l'Académie des Sciences". The results summarized here have been made possible thanks to the very successful operation of the LHC by the CERN accelerator departments. Thanks above all to colleagues from the technical and administrative staffs at CERN and at other ATLAS and CMS institutes for their contributions to the success of the experiments at the LHC.

References

[1] Particle Data Group, K. Nakamura et al., *Review of particle physics*, J. Phys. **G37** (2010) 075021.

[2] ATLAS Collaboration, G. Aad et al., *Observation of a new particle in the search for the Standard Model Higgs boson with the ATLAS detector at the LHC*, Phys. Lett. **B716** (2012) 1–29, arXiv:1207.7214 [hep-ex].

[3] CMS Collaboration, S. Chatrchyan et al., *Observation of a new boson at a mass of* 125 *GeV with the CMS experiment at the LHC*, Phys. Lett. **B716** (2012) 30–61, arXiv:1207.7235 [hep-ex].

[4] CMS Collaboration, S. Chatrchyan et al., *Observation of a new boson with mass near* 125 *GeV in pp collisions at* $\sqrt{s} = 7$ *and* 8 *TeV*, JHEP **1306** (2013) 081, arXiv:1303.4571 [hep-ex].

[5] S. Glashow, *Partial Symmetries of Weak Interactions*, Nucl. Phys. **22** (1961) 579–588.

[6] S. Weinberg, *A Model of Leptons*, Phys. Rev. Lett. **19** (1967) 1264–1266.

[7] A. Salam, *Weak and Electromagnetic Interactions*, Proceedings of the eighth Nobel symposium – N. Svartholm, ed. **C680519** (1968) 367–377.

[8] F. Englert and R. Brout, *Broken Symmetry and the Mass of Gauge Vector Mesons*, Phys. Rev. Lett. **13** (1964) 321–323.

[9] P.W. Higgs, *Broken symmetries, massless particles and gauge fields*, Phys. Lett. **12** (1964) 132–133.

[10] P.W. Higgs, *Broken Symmetries and the Masses of Gauge Bosons*, Phys. Rev. Lett. **13** (1964) 508–509.

[11] G. Guralnik, C. Hagen, and T. Kibble, *Global Conservation Laws and Massless Particles*, Phys. Rev. Lett. **13** (1964) 585–587.

[12] P.W. Higgs, *Spontaneous Symmetry Breakdown without Massless Bosons*, Phys. Rev. **145** (1966) 1156–1163.

[13] T. Kibble, *Symmetry breaking in non-Abelian gauge theories*, Phys. Rev. **155** (1967) 1554–1561.

[14] J.M. Cornwall, D.N. Levin, and G. Tiktopoulos, *Uniqueness of spontaneously broken gauge theories*, Phys. Rev. Lett. **30** (1973) 1268–1270.

[15] J.M. Cornwall, D.N. Levin, and G. Tiktopoulos, *Derivation of Gauge Invariance from High-Energy Unitarity Bounds on the s Matrix*, Phys. Rev. **D10** (1974) 1145.

[16] C. Llewellyn Smith, *High-Energy Behavior and Gauge Symmetry*, Phys. Lett. **B46** (1973) 233–236.

[17] B.W. Lee, C. Quigg, and H. Thacker, *Weak Interactions at Very High-Energies: The Role of the Higgs Boson Mass*, Phys. Rev. **D16** (1977) 1519.

[18] ATLAS Collaboration, G. Aad et al., *The ATLAS Experiment at the CERN Large Hadron Collider*, JINST **3** (2008) S08003.

[19] CMS Collaboration, S. Chatrchyan et al., *The CMS experiment at the CERN LHC*, JINST **3** (2008) S08004.

[20] LHC Higgs Cross Section Working Group, S. Heinemeyer et al., *Handbook of LHC Higgs Cross Sections: 3. Higgs Properties*, arXiv:1307.1347 [hep-ph].

[21] LEP Working Group for Higgs boson searches, ALEPH Collaboration, DELPHI Collaboration, L3 Collaboration, OPAL Collaboration, R. Barate et al., *Search for the standard model Higgs boson at LEP*, Phys. Lett. **B565** (2003) 61–75, arXiv:hep-ex/0306033 [hep-ex].

[22] TEVNPH Working Group, CDF Collaboration, D0 Collaboration, *Combined CDF and D0 Searches for the Standard Model Higgs Boson Decaying to Two Photons with up to 8.2 fb^{-1}*, arXiv:1107.4960 [hep-ex].

[23] M. Baak, M. Goebel, J. Haller, A. Hoecker, D. Ludwig, et al., *Updated Status of the Global Electroweak Fit and Constraints on New Physics*, Eur. Phys. J. **C72** (2012) 2003, arXiv:1107.0975 [hep-ph].

[24] CMS Collaboration, *Search for the Higgs Boson in the Fully Leptonic W^+W^- Final State*, CMS-PAS-HIG-11-014, http://cds.cern.ch/record/1376638.

[25] ATLAS Collaboration, *Search for the Standard Model Higgs boson in the $H \rightarrow WW^{(*)} \rightarrow \ell\nu\ell\nu$ decay mode ...*, ATLAS-CONF-2011-134, http://cds.cern.ch/record/1383837.

[26] ATLAS Collaboration, G. Aad et al., *Measurement of Higgs boson production in the diphoton decay channel in pp collisions at center-of-mass energies of 7 and 8 TeV with the ATLAS detector*, arXiv:1408.7084 [hep-ex].

[27] CMS Collaboration, V. Khachatryan et al., *Observation of the diphoton decay of the Higgs boson and measurement of its properties*, arXiv:1407.0558 [hep-ex].

[28] ATLAS Collaboration, G. Aad et al., *Measurement of the Higgs boson mass from the $H \rightarrow \gamma\gamma$ and $H \rightarrow ZZ^* \rightarrow 4\ell$ channels with the ATLAS detector using 25 fb^{-1} of pp collision data*, arXiv:1406.3827 [hep-ex].

[29] ATLAS Collaboration, G. Aad et al., *Measurements of Higgs boson production and couplings in diboson final states with the ATLAS detector at the LHC*, Phys. Lett. **B726** (2013) 88–119, arXiv:1307.1427 [hep-ex].

[30] CMS Collaboration, S. Chatrchyan et al., *Measurement of the properties of a Higgs boson in the four-lepton final state*, Phys. Rev. **D89** (2014) 092007, arXiv:1312.5353 [hep-ex].

[31] CMS Collaboration, *Precise determination of the mass of the Higgs boson and tests of compatibility of its couplings with the standard model predictions using proton collisions at 7 and 8 TeV*, CMS-HIG-14-009,CERN-PH-EP-2014-288, `arXiv:1412.8662`.

[32] ATLAS Collaboration, G. Aad et al., *Measurements of Higgs boson production and couplings in the four-lepton channel in pp collisions at center-of-mass energies of 7 and 8 TeV with the ATLAS detector*, `arXiv:1408.5191 [hep-ex]`.

[33] F. Caola and K. Melnikov, *Constraining the Higgs boson width with ZZ production at the LHC*, Phys. Rev. **D88** (2013) 054024, `arXiv:1307.4935 [hep-ph]`.

[34] N. Kauer and G. Passarino, *Inadequacy of zero-width approximation for a light Higgs boson signal*, JHEP **1208** (2012) 116, `arXiv:1206.4803 [hep-ph]`.

[35] N. Kauer, *Inadequacy of zero-width approximation for a light Higgs boson signal*, Mod. Phys. Lett. **A28** (2013) 1330015, `arXiv:1305.2092 [hep-ph]`.

[36] J.M. Campbell, R.K. Ellis, and C. Williams, *Bounding the Higgs width at the LHC using full analytic results for $gg \to e^-e^+\mu^-\mu^+$*, JHEP **1404** (2014) 060, `arXiv:1311.3589 [hep-ph]`.

[37] CMS Collaboration, V. Khachatryan et al., *Constraints on the Higgs boson width from off-shell production and decay to Z-boson pairs*, Phys. Lett. **B736** (2014) 64, `arXiv:1405.3455 [hep-ex]`.

[38] ATLAS Collaboration, *Determination of the off-shell Higgs boson signal strength in the high-mass ZZ final state with the ATLAS detector*, ATLAS-CONF-2014-042, `http://cds.cern.ch/record/1740973`.

[39] J.C. Collins and D.E. Soper, *Angular Distribution of Dileptons in High-Energy Hadron Collisions*, Phys. Rev. **D16** (1977) 2219.

[40] CMS Collaboration, S. Chatrchyan et al., *Study of the Mass and Spin-Parity of the Higgs Boson Candidate Via Its Decays to Z Boson Pairs*, Phys. Rev. Lett. **110** (2013) 081803, `arXiv:1212.6639 [hep-ex]`.

[41] CMS Collaboration, S. Chatrchyan et al., *Measurement of Higgs boson production and properties in the WW decay channel with leptonic final states*, JHEP **1401** (2014) 096, `arXiv:1312.1129 [hep-ex]`.

[42] ATLAS Collaboration, G. Aad et al., *Evidence for the spin-0 nature of the Higgs boson using ATLAS data*, Phys. Lett. **B726** (2013) 120–144, `arXiv:1307.1432 [hep-ex]`.

[43] CMS Collaboration, *Constraints on the spin-parity and anomalous HVV interactions of the Higgs boson from the CMS experiment*, CMS-PAS-HIG-14-018, `http://cds.cern.ch/record/1969386`.

[44] ATLAS Collaboration, G. Aad et al., *Search for the $b\bar{b}$ decay of the Standard Model Higgs boson in associated (W/Z)H production with the ATLAS detector*, `arXiv:1409.6212 [hep-ex]`.

[45] CMS Collaboration, S. Chatrchyan et al., *Search for the standard model Higgs boson produced in association with a W or a Z boson and decaying to bottom quarks*, Phys. Rev. **D89** (2014) 012003, `arXiv:1310.3687 [hep-ex]`.

[46] J.M. Butterworth, A.R. Davison, M. Rubin, and G.P. Salam, *Jet substructure as a new Higgs search channel at the LHC*, Phys. Rev. Lett. **100** (2008) 242001, `arXiv:0802.2470 [hep-ph]`.

[47] ATLAS Collaboration, *Evidence for Higgs Boson Decays to the* $\tau^+\tau^-$ *Final State with the ATLAS Detector*, ATLAS-CONF-2013-108, http://cds.cern.ch/record/1632191.

[48] CMS Collaboration, S. Chatrchyan et al., *Evidence for the 125 GeV Higgs boson decaying to a pair of τ leptons*, JHEP **1405** (2014) 104, arXiv:1401.5041 [hep-ex].

[49] CDF Collaboration, D0 Collaboration, T. Aaltonen et al., *Evidence for a particle produced in association with weak bosons and decaying to a bottom-antibottom quark pair in Higgs boson searches at the Tevatron*, Phys. Rev. Lett. **109** (2012) 071804, arXiv:1207.6436 [hep-ex].

[50] CDF Collaboration, D0 Collaboration, T. Aaltonen et al., *Higgs Boson Studies at the Tevatron*, Phys. Rev. **D88** (2013) no. 5, 052014, arXiv:1303.6346 [hep-ex].

[51] CMS Collaboration, S. Chatrchyan et al., *Evidence for the direct decay of the 125 GeV Higgs boson to fermions*, Nature Phys. **10** (2014) , arXiv:1401.6527 [hep-ex].

[52] ATLAS Collaboration, G. Aad et al., *Search for the Standard Model Higgs boson decay to* $\mu^+\mu^-$ *with the ATLAS detector*, Phys. Lett. **B738** (2014) 68–86, arXiv:1406.7663 [hep-ex].

[53] CMS Collaboration, V. Khachatryan et al., *Search for a standard model-like Higgs boson in the* $\mu^+\mu^-$ *and* e^+e^- *decay channels at the LHC*, arXiv:1410.6679 [hep-ex].

[54] ATLAS Collaboration, *Updated coupling measurements of the Higgs boson with the ATLAS detector using up to 25 fb^{-1} of proton-proton collision data*, ATLAS-CONF-2014-009, https://cds.cern.ch/record/1670012.

[55] CMS Collaboration, V. Khachatryan et al., *Search for the associated production of the Higgs boson with a top-quark pair*, JHEP **1409** (2014) 087, arXiv:1408.1682 [hep-ex].

Yves Sirois
LLR
École Polytechnique
CNRS-IN2P3
e-mail: yves.sirois@in2p3.fr

The H Boson, 51–64
© 2017 Springer Basel AG

Scalar Bosons and Supersymmetry

Pierre Fayet

Abstract. The recent discovery of a spin-0 Brout–Englert–Higgs boson leads to further enquire about other fundamental scalars. Supersymmetric theories involve, in relation with the electroweak breaking, five such scalars at least, two charged and three neutral ones, usually denoted as $H^\pm$, H, h and A. They also introduce spin-0 squarks and sleptons as the superpartners of quarks and leptons.

Supersymmetric extensions of the standard model lead to the possibility of gauge/BEH unification by providing spin-0 bosons as extra states for spin-1 gauge bosons within massive gauge multiplets. Depending on its properties the 125 GeV boson observed at CERN may then also be interpreted, up to a mixing angle induced by supersymmetry breaking, as *the spin-0 partner of the Z under two supersymmetry transformations*, i.e., as a Z that would be deprived of its spin.

1. The electroweak symmetry breaking

Special relativity and quantum mechanics, operating within quantum field theory, led to the Standard Model of particles and interactions. It has met a long series of successes with the discoveries of weak neutral currents (1973), charmed particles (1974–76), gluons mediators of strong interactions (1979), $W^\pm$ and Z's mediators of weak interactions (1983), and the sixth quark known as the top quark (1995).

Weak, electromagnetic and strong interactions are all understood from the exchanges of spin-1 mediators, $W^\pm$'s and Z's, photons and gluons, between spin-$\frac{1}{2}$ quarks and leptons, generically referred to as the constituents of matter. The u and d quarks are the building blocks for the protons uud and neutrons ddu, and the leptons include the electrons, muons and taus with their three neutrinos. The known fundamental particles are shown in Table 1, without any fundamental spin-0 boson yet.

The spin-1 bosons mediators of interactions are associated with local gauge symmetries. The eight gluons mediate the strong interactions, invariant under the color $SU(3)$ gauge group. The $W^\pm, Z$ and photon, mediators of the electroweak

TABLE 1. The gauge bosons mediators of strong, weak and electromagnetic interactions, and the quarks and leptons, constituents of matter.

spin-1 bosons *mediators of interactions*:   gluons, W^+, W^-, Z, photon,

spin-$\frac{1}{2}$ fermions *constituents of matter*:
$$\begin{cases} \text{6 quarks:} \ \begin{pmatrix} u \\ d \end{pmatrix} \begin{pmatrix} c \\ s \end{pmatrix} \begin{pmatrix} t \\ b \end{pmatrix}, \\ \text{6 leptons:} \ \begin{pmatrix} \nu_e \\ e^- \end{pmatrix} \begin{pmatrix} \nu_\mu \\ \mu^- \end{pmatrix} \begin{pmatrix} \nu_\tau \\ \tau^- \end{pmatrix}. \end{cases}$$

interactions, are associated with the $SU(2) \times U(1)$ electroweak gauge group [1–3]. It gives very much the same role to the left-handed quark fields u_L and d_L, and similarly to the left-handed lepton fields ν_{eL} and e_L.

The electroweak symmetry requires in principle the corresponding spin-1 gauge bosons to be massless. The weak interactions, mediated by virtual $W^\pm$ and Z production or exchanges, would then be long-ranged. They have instead a very short range $\simeq 2 \ 10^{-16}$ cm, corresponding to the large masses $m_W \simeq 80$ GeV/c^2 and $m_Z \simeq 91$ GeV/c^2 of their mediators, almost 100 times the mass of a proton. The electroweak symmetry should also require the charged leptons and quarks to be massless, which is not the case.

Both problems are solved within the standard model through the spontaneous breaking of the electroweak gauge symmetry induced by a doublet of complex spin-0 fields φ [3]. Three of its four real components, instead of being associated with three unwanted massless Goldstone bosons [4] as it would be the case if the electroweak $SU(2) \times U(1)$ symmetry were only global, are eliminated by the Brout–Englert–Higgs mechanism [5–7] to provide the additional degrees of freedom required for the $W^\pm$ and Z to acquire masses.

The fourth component of the spin-0 doublet, taken as $\phi = \sqrt{2\,\varphi^\dagger\varphi}$, adjusts uniformly in space-time so that the potential

$$V(\varphi) = \lambda \, (\varphi^\dagger\varphi)^2 - \mu^2 \, \varphi^\dagger\varphi, \tag{1}$$

with its famous mexican-hat shape, is minimum, for $\phi = v = \sqrt{\mu^2/\lambda}$ [3–6]. The electroweak symmetry is then said to be "spontaneously broken" (even if ϕ itself remains gauge-invariant), meaning by this expression that the $W^\pm$ and Z are no longer massless. The local gauge symmetry, which strictly speaking still remains unbroken, gets now hidden. The $W^\pm$ and Z acquire masses fixed in terms of the electroweak gauge couplings g and g' by

$$m_W = \frac{gv}{2}, \quad m_Z = \frac{\sqrt{g^2 + g'^2}\ v}{2} = \frac{m_W}{\cos\theta}. \tag{2}$$

The electroweak mixing angle θ which enters in the definitions of the Z and photon fields is fixed by $\tan\theta = g'/g$, the photon staying massless.

This mechanism of spontaneous symmetry breaking is at the origin of the differentiation between weak interactions, becoming short-ranged, and electromagnetic ones, which remain long-ranged. The elementary charge e and the Fermi coupling of weak interactions G_F are given by $e = g \sin\theta$ and $G_F/\sqrt{2} = g^2/(8m_W^2) = 1/(2v^2)$, so that $v = (G_F\sqrt{2})^{-1/2} \simeq 246$ GeV.

TABLE 2. Particle content of the standard model. Strong, weak and electromagnetic interactions of quarks and leptons are invariant under the $SU(3) \times SU(2) \times U(1)$ gauge symmetry group. The spin-0 BEH boson is associated with the spontaneous breaking of the electroweak symmetry and the generation of masses for the $W^{\pm}$ and Z, quarks and charged leptons.

spin-1 gauge bosons:	gluons, W^+, W^-, Z, photon
spin-$\frac{1}{2}$ fermions:	6 quarks + 6 leptons
1 spin-0	scalar BEH boson

The introduction of the spin-0 doublet field φ, called a Brout–Englert–Higgs field or often simply Higgs field, also allows for the charged leptons and quarks to acquire masses, that would otherwise be forbidden by the $SU(2) \times U(1)$ symmetry. Indeed an electron in an electromagnetic field acquires an energy $E = qV$ where $q = -e$ is its electric charge and V the electrostatic potential. The electron field, which interacts with the spin-1 electromagnetic gauge field A^μ, also interacts with the spin-0 doublet field φ, with a coupling constant λ_e. The electron is then sensitive to its modulus, i.e., to the physical BEH field (still gauge invariant),

$$\phi = \sqrt{2\,\varphi^\dagger\varphi}\,. \tag{3}$$

It acquires a field-dependent mass parameter $\lambda_e\phi$ and thus, with the BEH field having a constant value $\phi = v$ uniform in space-time, a mass $m_e = \lambda_e v$. The same phenomenon occurs for the other charged leptons μ and τ, and the quarks. They all acquire masses proportional to their couplings to this scalar field ϕ,

$$m_l = \lambda_l\,v\,, \quad m_q = \lambda_q\,v\,. \tag{4}$$

The three neutrinos, which do not interact directly with ϕ in connection with the absence of right-handed neutrino fields ν_R, remain massless at this stage. They have, however, very small masses whose origin is as yet unknown, and can oscillate from one flavor to another.

2. The scalar boson of the standard model

An immediate consequence of the introduction of a spin-0 field φ is that there should be spin-0 excitations, i.e., particles, associated with its quantization. The complex electroweak doublet φ introduced [3] to generate a spontaneous breaking of the electroweak symmetry [1,2] describes four field degrees of freedom. Three of them are eliminated by the BEH mechanism [5–7], with one only (indeed the one which attracted the least attention at the time) surviving in the physical theory.

The waves corresponding to the space-time variations of this field $\phi = \sqrt{2\varphi^\dagger\varphi}$, once quantized, are associated with neutral scalar Brout–Englert–Higgs bosons, also more commonly referred to as Higgs bosons for historical reasons. This is very much the same as for electromagnetic waves, whose quanta are the massless spin-1 photons. The mass of this spin-0 boson, obtained by expanding the potential (1) near its minimum, is given by

$$m_h = \sqrt{2\mu^2} = \sqrt{2\lambda v^2}. \tag{5}$$

This scalar is essential for the consistency of the standard model as a quantum field theory. Its couplings to quarks and leptons, obtained from (4), are proportional to their masses,

$$\lambda_{q,l} = 2^{1/4} G_F^{1/2} m_{q,l}. \tag{6}$$

Its mass, however, is not predicted by the theory, but fixed by the quartic coupling λ in the scalar potential $V(\varphi)$ in (1).

The possible origin of this $\lambda (\varphi^\dagger\varphi)^2$ quartic coupling is a subject to which we shall return later, within the framework of supersymmetric theories. They lead to consider several spin-0 BEH bosons, charged and neutral, relating their quartic couplings to the squares of the electroweak gauge couplings, g^2 and g'^2, in particular through

$$\lambda = \frac{g^2 + g'^2}{8}. \tag{7}$$

They thus also provide, in particular, a neutral spin-0 BEH boson that would have the same mass as the Z [8],

$$m_h = \sqrt{2\lambda v^2} = \frac{\sqrt{g^2 + g'^2}\, v}{2} = m_Z \simeq 91 \text{ GeV}/c^2, \tag{8}$$

in the absence of supersymmetry breaking effects.

The scalar boson of the standard model has long remained its last missing particle after the discovery of the top quark in 1995, escaping until recently all experimental efforts deployed to detect its production, most notably at the e^+e^- LEP collider at CERN, which established a lower bound of 114 GeV/c^2 on its mass [9].

The existence of a new boson has been established recently, in 2012, at the Large Hadron Collider LHC at CERN [10,11]. This particle, neutral, has a mass close to 125 GeV/c^2, and almost certainly spin 0, rather than 2. It cannot have spin 1 as it is observed to have $\gamma\gamma$ decay modes. It shows at this point the properties expected from a scalar boson associated with the differentiation between electro-magnetic and weak interactions, and the generation of masses. If it is indeed the scalar boson of the standard model this one may now be considered as complete, this spin-0 particle being its last missing piece (cf. Table 2). The standard model would then become the standard theory of particle interactions.

3. A scalar boson, elementary or not?

The existence of such a scalar boson has in fact long been questioned, many physi-cists having serious doubts about the very existence of fundamental spin-0 fields. Indeed in a theory including very high scales much larger than the electroweak scale, such as a possible grand-unification scale (now believed to be of the order of 10^{16} GeV), or the Planck scale $\simeq 10^{19}$ GeV possibly associated with quantum gravity, such spin-0 fields tend to acquire very large mass terms, disappearing from the low-energy theory.

Many efforts were thus devoted to replace fundamental spin-0 fields, with-out much success, by composite fields built from spin-$\frac{1}{2}$ ones, e.g., techniquark fields specially introduced for that purpose [12–15], in view of ultimately avoiding fundamental spin-0 bosons associated with the electroweak breaking.

One may ask whether the new boson recently found at CERN [10, 11] is indeed the scalar one of the standard model, or if its properties may deviate from it at some point. It is of course not the first spin-0 particle found. Pions, kaons, ... are also spin-0 particles, but composite $q\bar{q}$ states constructed from quarks and antiquarks. In contrast the new 125 GeV boson presents at this stage all the characteristics of an elementary particle, the first of its kind. Is it alone, or just the first member of a new class?

4. Introducing supersymmetry

In the meantime however, the situation concerning our view of spin-0 fields has changed with the introduction of supersymmetry. Its algebraic structure [16–21] provides a natural framework for fundamental spin-0 fields, now treated on the same footing as spin-$\frac{1}{2}$ ones, to which they are related by supersymmetry trans-formations. Indeed, according to common knowledge, supersymmetry is expected to relate bosons (of integer spin) with fermions (of half-integer spin), as follows:

$$\begin{array}{ccc} & \text{supersymmetry} & \\ \text{bosons} & \longleftrightarrow & \text{fermions}\,. \\ \text{(integer spin)} & & \text{(half-integer spin)} \end{array} \qquad (9)$$

But can this be of any help in understanding the real world of particles and interactions? If supersymmetry is to act at the fundamental level the natural idea would be to try to use it to relate the known bosons and fermions in Table 1, or 2. More precisely, can one relate the spin-1 bosons (gluons, $W^\pm$, Z and photon) messengers of interactions to the spin-$\frac{1}{2}$ fermions (quarks and leptons) constituents of matter? This would lead to a sort of unification

$$\text{Forces} \quad \overset{\text{supersymmetry?}}{\longleftrightarrow} \quad \text{Matter}. \tag{10}$$

The idea looks attractive, even so attractive that supersymmetry is frequently presented as a symmetry uniting forces with matter. This is however misleading at least at the present stage, and things do not work out that way.

Indeed the algebraic structure of supersymmetry did not seem directly applicable to particle physics [21], in particular as known fundamental bosons and fermions do not seem to have much in common. There are also other more technical reasons, dealing with the difficulties of spontaneous supersymmetry breaking, the fate of the resulting Goldstone fermion [22, 23] (even if it is subsequently eaten away by the spin-$\frac{3}{2}$ gravitino [25]), the presence of self-conjugate Majorana fermions, the requirements of baryon and lepton number conservation, etc..

5. Relating bosons and fermions, yes, but how?

One has to find out which bosons and fermions might be related under supersymmetry, first considering possible associations between baryons and mesons. Or, at the fundamental level, exploring tentative associations like

$$\left\{ \begin{array}{ccc} \text{photon} & \overset{?}{\longleftrightarrow} & \text{neutrino} \\ W^{\pm} & \overset{?}{\longleftrightarrow} & e^{\pm} \\ \text{gluons} & \overset{?}{\longleftrightarrow} & \text{quarks} \\ & \cdots & \end{array} \right. \tag{11}$$

But we have no chance to realize in this way systematic associations of known fundamental bosons and fermions. This is also made obvious from the fact that we know 90 fermionic field degrees of freedom for the quarks and leptons (for 3 families of 15 chiral quark and lepton fields) as compared to 28 only for bosonic ones (16 + 11 + 1 for the gluons, electroweak gauge bosons and the new scalar). In addition these fields have different gauge and B and L quantum numbers, preventing them from being directly related.

In supersymmetry we also have to deal with the systematic appearance of self-conjugate Majorana fermions, while Nature seems to know Dirac fermions only (with a possible special exception for neutrinos having Majorana mass terms). How can we obtain Dirac fermions, and attribute them conserved quantum numbers like B and L?

6. The need for superpartners

To face all these difficulties, we were led to introduce a color octet of spin-$\frac{1}{2}$ Majorana fermions [24] called *gluinos* [25], although their consideration was at the time forbidden by the general principle of triality [26]. This one, however, gets systematically violated within supersymmetric theories, in which it no longer applies. A strict application of this principle would have prevented us from discussing supersymmetric theories.

As gluons were associated with gluinos, the photon had to be associated, not with any the known neutrinos ν_e and ν_μ, or later ν_τ, but with a "photonic neutrino" called the *photino*. It is the first member of a larger family of *neutralinos*, obtained from mixings of neutral spin-$\frac{1}{2}$ gaugino and higgsino fields, respectively associated with gauge and BEH fields under supersymmetry.

But, as far as we know at the moment, baryon and lepton numbers B and L are carried by fundamental fermions only, quarks and leptons, not by bosons. They are thus even referred to as "fermionic numbers". In a supersymmetric theory however it gets impossible to have B and L carried just by fermions, and not by bosons. But attributing "fermion number" to bosons looks like a non-sense! To include the standard model within a supersymmetric theory we also have to accept the unconventional idea that a significant number of fundamental bosons may have to carry baryon or lepton numbers.

These new bosons carrying B or L are now well known as *squarks* and *sleptons*. Their denomination makes apparent and even "obvious" that they have to carry the same B and L as their fermionic counterparts.

Still this does not guarantee yet that B and L will systematically remain conserved as observed in Nature, at least to a sufficiently good approximation. In particular we do not want the proton to undergo too fast decays, as its lifetime should be larger than about 10^{32} years or so. This requires the consideration of an additional symmetry, namely R symmetry or its discrete version, R-parity [8, 24, 27].

This one, originally obtained as the parity of a continuous quantum number R carried by the supersymmetry generator, $R_p = (-1)^R$, is simply +1 for all standard model particles and -1 for their superpartners,

$$
R_p = \begin{cases} +1 & \text{for } quarks\ and\ leptons,\ gauge\ and\ BEH\ bosons, \\ -1 & \text{for } squarks\ and\ sleptons,\ gluinos,\ charginos\ and\ neutralinos. \end{cases}
$$
$$(12)$$

It may be rewritten in terms of the spin, baryon and lepton numbers of particles, as

$$
R_p = (-1)^{2S}\,(-1)^{3B+L} .
$$
$$(13)$$

This illustrates its connection with B and L, or simply $B - L$, conservation laws, even allowing for $\Delta L = \pm 2$ processes as in the presence of neutrino Majorana mass terms.

Although one may consider that R-parity conservation is not necessary and may thus be questioned [29], its requirement is natural, and its absence usually the source of various troubles. Indeed R-parity prevents direct exchanges of squarks or sleptons between ordinary quarks and leptons, that could induce proton decay at a much too high rate.

R-parity also plays a crucial role in the stability of the lightest supersymmetric particle, or LSP, in general considered to be a *neutralino*. The pair-production (and subsequent decays) of supersymmetric particles should ultimately lead to two unobserved neutralinos, the famous missing energy-momentum signature often used to search for supersymmetry. Stable massive neutralinos having survived annihilations also turn out to be natural candidates for the non-baryonic dark matter of the Universe.

Supersymmetry thus does not relate directly known bosons and fermions. All known particles should be associated with new superpartners which appear as their *images under supersymmetry*, according to

$$\left\{ \begin{array}{lcl} \text{known bosons} & \longleftrightarrow & \text{new fermions,} \\ \text{known fermions} & \longleftrightarrow & \text{new bosons.} \end{array} \right. \tag{14}$$

This was long mocked as a sign of the irrelevance of supersymmetry. But times have changed, to the point that supersymmetry gets now frequently referred to as a symmetry which postulates the existence, for each particle of the standard model, of a supersymmetric partner differing by $1/2$ unit of spin.

7. The supersymmetric standard model, with its extra spin-0 bosons

The resulting supersymmetric standard model involves spin-0 squarks and sleptons, and spin-$\frac{1}{2}$ gluinos, charginos (called winos in Table 3) and neutralinos [8, 24, 27, 28]. As of today, however, they remain unseen after more than three decades of experimental searches, starting in the late seventies. The search for these new particles is now one of the main objectives of the LHC collider at CERN [30, 31], whose energy should soon increase from 8 to 13 TeV.

Will this be sufficient? At which energy scale should the new supersymmetric particles be found? Is it of the order of the TeV scale, not too far from the electroweak scale and accessible at LHC? Or possibly significantly larger, as it could happen in theories with extra space dimensions [32, 33], with the mass scale of the new superpartners fixed by the (or a) compactification scale $\propto \hbar/Lc$? ($L < 10^{-17}$ cm corresponding to $\hbar c/L > 2$ TeV.)

In any case the supersymmetric standard model requires for the electroweak breaking, not a single doublet φ as in the standard model, but two at least,

$$h_1 = \begin{pmatrix} h_1^0 \\ h_1^- \end{pmatrix} \quad \text{and} \quad h_2 = \begin{pmatrix} h_2^+ \\ h_2^0 \end{pmatrix}. \tag{15}$$

They are needed to construct two massive Dirac charginos (also called winos in Table 3) from charged gaugino ($\widetilde{W}_{L+R}^-$) and higgsino ($\tilde{h}_{1L}^-$ and $(\tilde{h}_{2L}^+)^c$) components, without getting stuck with a massless chiral charged fermion [8, 24]. These

TABLE 3. Minimal content of the Supersymmetric Standard Model (MSSM). Neutral gauginos and higgsinos mix into a photino, two zinos and a higgsino, further mixed into four neutralinos. Ordinary particles from the standard model, including additional BEH bosons, in blue, have R-parity $+1$. Their superpartners, in red, have R-parity -1.

Spin 1	Spin 1/2	Spin 0
gluons photon	gluinos $\tilde{g}$ photino $\tilde{\gamma}$	
$W^{\pm}$ Z	winos $\widetilde{W}_{1,2}^{\pm}$ zinos $\widetilde{Z}_{1,2}$ higgsino $\tilde{h}$	$H^{\pm}$ h $\Big\}$ BEH bosons $H,\ A$
	leptons l quarks q	sleptons $\tilde{l}$ squarks $\tilde{q}$

two doublets allow for the generation of charged-lepton and down-quark masses from h_1, and up-quark masses from h_2. This applies to the various versions of the supersymmetric standard model, from the minimal one known as the MSSM, to others that may include an extra singlet coupled to the two doublets h_1 and h_2 as in the N/nMSSM, or USSM if an extra $U(1)$ symmetry is gauged, GMSB models, etc.

h_1 and h_2 represent altogether eight real spin-0 fields, among which three get eliminated when the $W^{\pm}$ and Z acquire masses. This results in five spin-0 BEH scalars, two charged and three neutrals, usually denoted as $H^{\pm}$, H, h and A, also actively searched for at LHC [34, 35]. Additional ones may also exist beyond the MSSM, whose particle content is represented in Table 3.

8. Some remaining questions, after the 125 GeV boson discovery

The standard model constitues a remarkable achievement in the description of the fundamental particle interactions. Even if it is complete, it still leaves many questions unanswered. In addition, it would be presumptuous to imagine that our knowledge of particles and interactions is now complete, without new particles or interactions remaining to be discovered.

The standard model does not answer many questions, concerning the origin of symmetries and symmetry breaking, the quark and lepton mass spectrum and mixing angles, etc.. Gravitation, classically described by general relativity, cannot easily be cast into a consistent quantum theory. This is why string theories were developed, which seem to require supersymmetry for consistency. The nature of dark matter and dark energy which govern the evolution of the Universe and its accelerated expansion remains unknown, as the origin of the predominance of matter over antimatter.

Dark matter may be composed, for its main part, non-baryonic, of new particles such as the neutralinos of supersymmetric theories, or axions, ... There may also be new forces or interactions beyond the four known ones, strong, electromagnetic, weak and gravitational. And maybe, beyond space and time, new hidden dimensions, extremely small (with $L < 10^{-17}$ cm corresponding to $\hbar c/L > 2$ TeV) or even stranger, like the anticommuting dimensions of supersymmetry.

9. More on supersymmetry and superspace

Supersymmetry enlarges the notions of space and time, already related by the theory of relativity, to a new geometry involving a superspace [19, 36, 37]. This one possesses new quantum coordinates θ_α associated with rotations and Lorentz transformations. In the simplest case the θ_α's are the four components of a self-conjugate Majorana spinor θ called the Grassmann coordinate. They satisfy anti-commutation relations,

$$\theta_\alpha \theta_\beta = - \theta_\beta \theta_\alpha \,, \tag{16}$$

in connection with the spin-$\frac{1}{2}$ character of the Grassmann coordinate θ, and associated fermionic character of spin-$\frac{1}{2}$ particles obeying the Pauli exclusion principle. Each θ_α has a vanishing square, $\theta_\alpha^2 = 0$, very much like two identical spin-$\frac{1}{2}$ fermions cannot be in the same quantum state.

The notion of point gets replaced by the notion of "superspace point",

$$\text{space-time point } x^\mu = \begin{pmatrix} ct \\ \vec{x} \end{pmatrix} \quad \rightarrow \quad (x^\mu, \theta_\alpha)\,. \tag{17}$$

This means in fact that we shall now consider, in the place of fields $\varphi(x^\mu)$ depending on the time and space coordinates t and $\vec{x}$, superfields Φ depending on x^μ and θ_α. It is in practice convenient to rewrite θ as a 2-component complex spinor rather than a 4-component self-conjugate one. A superfield is then expressed as

$$\Phi(x, \theta, \bar{\theta})\,. \tag{18}$$

Its expansion in terms of θ and $\bar{\theta}$ provides a finite number of component fields, both bosonic and fermionic, in equal numbers. There are different types of superfields. Gauge superfields describe massless spin-1 gauge bosons and associated spin-$\frac{1}{2}$ gauginos. Chiral superfields describe both spin-$\frac{1}{2}$ and spin-0 fields, corresponding to quarks and leptons with associated squarks and sleptons for $R_p = -1$ superfields. Or to spin-0 BEH fields with associated higgsinos, for $R_p = +1$ superfields, with the θ coordinate having $R_p = -1$.

Supersymmetry transformations act in superspace in a special way reminiscent of a translation for the Grassmann coordinate θ, combined with a transformation of the space-time coordinate x^μ involving θ. Their generators Q_α are in the simplest case the four hermitian components of a self-conjugate Majorana spinor Q (or of a 2-component complex spinor). In a schematic way supersymmetry transformations generated by the Q_α's transform, for each superfield, bosonic components into fermionic ones, and conversely. Being of fermionic nature the operators Q_α satisfy anticommutation relations.

The combination of two infinitesimal rotations around Ox and Oy generates an infinitesimal rotation around the third orthogonal axis Oz, as expressed by the commutation relation

$$[J_x, J_y] = i\hbar J_z, \tag{19}$$

essential in quantum mechanics. In a somewhat similar way, the appropriate combination, now through the anticommutators $Q_\alpha Q_\beta + Q_\beta Q_\alpha$, of two infinitesimal supersymmetry transformations generates a translation in space-time. This is expressed by the following (anti)commutation relations in the supersymmetry algebra

$$\begin{cases} \{Q, \bar{Q}\} = -2\gamma_\mu P^\mu, \\ [Q, P^\mu] = 0. \end{cases} \tag{20}$$

The last commutation relation, $[Q, P^\mu] = 0$, expresses that supersymmetry transformations commute with translations.

Supersymmetry transformations generated by the spin-$\frac{1}{2}$ operator Q_α change the intrinsic angular momentum of particles, i.e., their spin, by half a unit (or $\hbar/2$). When local supersymmetry transformations are considered this necessitates the possibility of performing local space-time translations, with parameters $\epsilon^\mu(x)$. It then requires general relativity, and thus gravitation, to be included in the game, leading to theories of supergravity [38, 39], with the spin-2 graviton having for superpartner a spin-$\frac{3}{2}$ gravitino.

Although the gravitino is a priori coupled with a very small coupling constant $\kappa = \sqrt{8\pi G_N}$, of gravitational strength, it could have a very small mass $m_{3/2}$, depending on the models considered, as in the so-called GMSB models. It could then still play an important role in particle physics experiments [25], possibly appearing as the lightest supersymmetric particle produced at the end of a decay chain, carrying away missing energy-momentum.

10. Relating gauge and BEH bosons, in the supersymmetric standard model

In supersymmetric extensions of the standard model, the quartic interactions between the two doublets h_1 and h_2 appear as part of the electroweak gauge interactions, their quartic contribution to the potential providing the one of the MSSM [8],

$$V_{\text{quartic}}(h_1, h_2) = \frac{g^2 + g'^2}{8} (h_1^\dagger h_1 - h_2^\dagger h_2)^2 + \frac{g^2}{2} |h_1^\dagger h_2|^2. \tag{21}$$

The quartic coupling constants are no longer arbitrary as for λ in the standard model, but fixed in terms of the electroweak gauge couplings as $(g^2 + g'^2)/8$ and $g^2/2$. This is at the origin of the mass equality (8), for conserved supersymmetry.

This is also at the origin of a mass inequality requiring in the MSSM, at the classical level, the lightest spin-0 boson h to be lighter than $m_Z \simeq 91$ GeV/c^2, up to radiative corrections which could raise it up to 125 GeV/c^2 from large supersymmetry-breaking effects involving heavy stop quarks [20]. In the presence of an extra singlet coupled to h_1 and h_2, however, as in the N/nMSSM, extra contributions to this lightest mass make it easier to reach 125 GeV/c^2.

When the BEH mechanism operates within a supersymmetric theory, it provides *massive gauge multiplets* [8]. Each of them describes a massive spin-1 gauge boson, two spin-$\frac{1}{2}$ inos constructed from gaugino and higgsino components, and a spin-0 boson. The latter is actually a BEH boson associated with the spontaneous breaking of the gauge symmetry. We thus get systematic associations between massive gauge bosons and spin-0 BEH bosons, a quite non-trivial feature owing to their different gauge symmetry properties and very different couplings to quarks and leptons [33, 40], with the general association

$$Z \overset{SUSY}{\longleftrightarrow} 2 \text{ Majorana zinos } \overset{SUSY}{\longleftrightarrow} \text{ spin-0 BEH boson}. \qquad (22)$$

The spin-0 field partner of the Z, expressed in usual notations as

$$z = \sqrt{2} \text{ Re } (- h_1^0 \cos\beta + h_2^0 \sin\beta), \qquad (23)$$

may also be described in a non-conventional way by the massive Z gauge superfield, expanded as $Z(x, \theta, \bar{\theta}) = -z/m_Z + \cdots - \theta\sigma_\mu\bar{\theta} \, Z^\mu + \cdots$.

This implies the existence of a spin-0 BEH boson of mass

$$m \simeq 91 \text{ GeV}/c^2, \quad \text{up to supersymmetry-breaking effects.} \qquad (24)$$

More precisely the z field (23) corresponds in the MSSM to a mixing between the lighter and heavier spin-0 eigenstates h and H. The spin-0 partner of the Z may then be identified, in the MSSM or N/nMSSM, ..., as the new 125 GeV boson, up to a mixing angle, possibly small, induced by supersymmetry breaking [40]. We also have in a similar way

$$W^\pm \overset{SUSY}{\longleftrightarrow} 2 \text{ Dirac winos } \overset{SUSY}{\longleftrightarrow} \text{ spin-0 boson } H^\pm, \qquad (25)$$

with $m_{H^\pm} = m_{W^\pm}$ up to supersymmetry-breaking effects.

11. Conclusion on scalar bosons and supersymmetry

In addition to superpartners, supersymmetric theories lead to an extended set of spin-0 bosons $H^\pm, H, h, A, \ldots$. Some appear as extra states for massive spin-1 gauge bosons, providing a relation between spin-1 mediators of gauge interactions and spin-0 particles associated with symmetry breaking and mass generation.

Depending on its properties the 125 GeV boson observed at CERN may also be interpreted, up to a mixing angle induced by supersymmetry breaking, as *the spin-0 partner of the Z under two supersymmetry transformations*, i.e., as a Z that would be deprived of its spin. This provides within a theory of electroweak and strong interactions the first example of two known fundamental particles of different spins that may be related by supersymmetry, in spite of their different electroweak properties.

The next run of LHC experiments may well allow for the direct production of supersymmetric particles. Even this does not happen, and R-odd superpartners were to remain out of reach for some time, possibly due to large momenta along very small compact dimensions, supersymmetry could still be tested in the gauge-and-BEH sector at present and future colliders, in particular through the properties of the new spin-0 boson.

References

[1] Glashow, S.: Nucl. Phys. **22**, 579 (1961).

[2] Salam, A., and Ward, J.: Phys. Lett. **13**, 168 (1964).

[3] Weinberg, S.: Phys. Rev. Lett. **19**, 1264 (1967).

[4] Goldstone, J.: Nuovo Cim. **19**, 154 (1961).

[5] Englert, F., and Brout, R.: Phys. Rev. Lett. **13**, 321 (1964).

[6] Higgs, P.: Phys. Rev. Lett. **13**, 508 (1964).

[7] Guralnik, G., Hagen, C., and Kibble, T.: Phys. Rev. Lett. **13**, 585 (1964).

[8] Fayet, P.: Nucl. Phys. B **90**, 104 (1975).

[9] ALEPH, DELPHI, L3 and OPAL coll.: Phys. Lett. B **565**, 61 (2003).

[10] ATLAS coll.: Phys. Lett. B **716**, 1 (2012).

[11] CMS coll.: Phys. Lett. B **716**, 30 (2012).

[12] Weinberg, S.: Phys. Rev. D **13**, 974 (1976); D **19**, 1277 (1979).

[13] Susskind, L.: Phys. Rev. D **20**, 2619 (1979).

[14] Dimopoulos, S., and Susskind, L.: Nucl. Phys. B **155**, 237 (1979).

[15] Eichten, E., and Lane, K.: Phys. Lett. B **90**, 125 (1980).

[16] Gol'fand, Yu., and Likhtman, E.: ZhETF Pis. Red. **13**, 452 (1971) [JETP Lett. **13**, 323 (1971)].

[17] Volkov, D., and Akulov, V.: Phys. Lett. B **46**, 109 (1973).

[18] Wess, J., and Zumino, B.: Nucl. Phys. B **70**, 39 (1974).

[19] Fayet, P., and Ferrara, S.: Phys. Rep. **32**, 249 (1977).

[20] Martin, S.: A Supersymmetry Primer, http://arxiv.org/abs/hep-ph/9709356 (2011).

[21] Ramond, P.: Eur. Phys. J. C **74**, 2698 (2014).

[22] Fayet, P., and Iliopoulos, J.: Phys. Lett. B **51**, 461 (1974).

[23] Fayet, P.: Phys. Lett. B **58**, 67 (1975); O'Raifeartaigh, L.: Nucl. Phys. B **96**, 331 (1975).

[24] Fayet, P.: Phys. Lett. B **64**, 159 (1976); B **69**, 489 (1977).

[25] Fayet, P.: Phys. Lett. B **70**, 461 (1977).

[26] Gell-Mann, M., Ramond, P., and Slansky, R.: Rev. Mod. Phys. **50**, 721 (1978).

[27] Farrar, G., and Fayet, P.: Phys. Lett. B **76**, 575 (1978).

[28] Fayet, P.: Phys. Lett. B **84**, 416 (1979).

[29] Barbier, R., *et al.*: Phys. Rep. **420**, 1 (2005).

[30] https://twiki.cern.ch/twiki/bin/view/AtlasPublic/
     SupersymmetryPublicResults

[31] https://twiki.cern.ch/twiki/bin/view/CMSPublic/PhysicsResultsSUS

[32] Fayet, P.: Phys. Lett. B **159**, 121 (1985); Nucl. Phys. B **263**, 649 (1986).

[33] Fayet, P.: Eur. Phys. J. C **74**, 2837 (2014).

[34] https://twiki.cern.ch/twiki/bin/view/AtlasPublic/HiggsPublicResults

[35] https://twiki.cern.ch/twiki/bin/view/CMSPublic/PhysicsResultsHIG

[36] Salam, A., and Strathdee, J.: Nucl. Phys. B **76**, 477 (1974).

[37] Ferrara, S., Wess, J., and Zumino, B.: Phys. Lett. B **51**, 239 (1974).

[38] Ferrara, S., Freedman, D., and van Nieuwenhuizen, P.: Phys. Rev. D **13**, 3214 (1976).

[39] Deser, S., and Zumino, B.: Phys. Lett. B **62**, 335 (1976).

[40] Fayet, P.: Nucl. Phys. B **237**, 367 (1984); Phys. Rev. D **90**, 015033 (2014).

Pierre Fayet
Laboratoire de Physique Théorique
de l'École Normale Supérieure
24, rue Lhomond
F-75231 Paris cedex 05, France
    and
Département de Physique
École Polytechnique
F-91128 Palaiseau cedex, France
e-mail: pierre.fayet@lpt.ens.fr

The H Boson, 65–83
© 2017 Springer Basel AG

Future Searches on Scalar Boson(s)

Louis Fayard

Abstract. The discovery of a Brout–Englert–Higgs boson in July 2012 opened the road to new searches in the electroweak symmetry breaking sector, either doing precise measurements of the discovered boson, or looking for additional bosons. This can be done, either at the LHC or at future accelerators that will be described.

1. Introduction

The discovery in July 2012 [1, 2] of a (Brout–Englert–Higgs) BEH boson at the Large Hadron Collider (LHC) [3, 4] ended an era of search of the last piece of the Standard Model, culminating with the Nobel prize given to Englert and Higgs [5, 6] (detailed historical account can be found in [7, 8, 9]). However, even if the Standard Model (SM) agrees with all the experimental data from colliders, it is theoretically unsatisfactory (the hierarchy problem, too many parameters in the Model, ...) and some experimental observations cannot be accomodated (neutrino masses and oscillations, baryon asymmetry in the Universe, dark matter, ...). However no new physics was discovered at the LHC in the run 1 (mainly at a centre of mass energy of 8 TeV). This included Supersymmetry, which is the most popular extension able to accomodate a BEH boson mass close to the one measured [10]. This puts even more pressure to the next runs at the LHC, at a centre of mass energy of 13 or 14 TeV and to the design of the future accelerators. They will be described in the next section and the third section will discuss new physics in the scalar sector, either using the (already discovered) BEH boson as a probe for new physics or trying to find new scalar bosons.

2. Future facilities

The LHC 'short-term' upgrades will be described first, in Subsections 2.1 and 2.2. Then longer term future facilities will be described after. However muon colliders [11], photon colliders [12] and plasma-based particle acceleration [13] will not be described here. General reviews on future facilities can be found elsewhere [14, 15]. One can also have a look at the European Strategy [16] and P5 [17] reports or at recent conferences like *Higgs Hunting* [18] or recent ICFA workshops [19].

2.1. Runs 2 and 3 of LHC

The LHC will restart in the middle of 2015 at a centre of mass energy of 13 TeV (and maybe later 14 TeV) and will work during run 2 and run 3 (up to about 2022) at a luminosity close to $\sim 2\ 10^{34}$ cm^{-2} s^{-1}. The run 2 will last about 3 years with $\sim 50\ fb^{-1}$ each year and will be followed after a long shutdown (LS2) of $\sim$ one year by the run 3 with $\sim 60\ fb^{-1}$ each year [20].

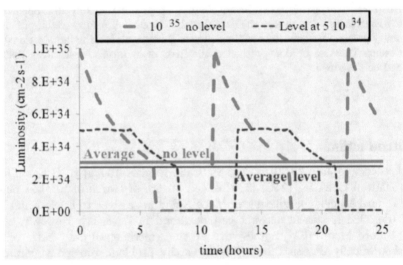

FIGURE 1. Luminosity levelling principle [11].

2.2. High-Luminosity LHC (HL-LHC)

After 10 years of operation, the performance of the LHC in terms of integrated luminosity will saturate. This is the main motivation to propose a High-Luminosity LHC (HL-LHC) [21, 22, 23], starting around 2025 and aiming at accumulate every year of HL-LHC the same total luminosity that could be obtained in the previous decade, i.e., 300 fb^{-1}. The guidelines are to have an accelerator able to reach $2\ 10^{35} cm^{-2} s^{-1}$, but to run in the initial part of the fill by luminosity levelling in the initial part of the fill at $5\ 10^{34} cm^{-2} s^{-1}$ (see Figure 1). This allows one to reach the integrated luminosity without having too large pile-up in the experiments at the beginning of the run. The main modifications, in addition to the new injector LINAC4 that should be connected in 2019 are

- Crab cavities to take advantage of the small β^*

- Cryo-collimators and high field (Nb_3Sn) 11 T dipoles in dispersion suppressors (at least close to ALICE)
- New high field (Nb_3Sn)/larger aperture interaction region magnets (in ATLAS and CMS. The current triplet assembly (mainly the orbit corrector magnets) will reach the end of their lifetime due to radiation damage at an integrated luminosity of ca. 300 fb^{-1}.

One should note that some 11 T (Nb_3Sn) dipoles have already been tested at Fermilab. Additional information about HL-LHC can be found in the recent ECFA workshop [24].

2.3. Linear Colliders

The Linear Collider Collaboration [25] is an organisation that brings the two most likely candidates, the Compact Linear Collider Study (CLIC) and the International Linear Collider (ILC), together under one roof.

2.3.1. The International Linear Collider (ILC). The International Linear Collider (ILC) is a high-luminosity linear electron-positron collider based on 1.3 GHz superconducting radio-frequency (SCRF) accelerating technology. Its centre-of-mass-energy range is 200 $\sim$ 500 GeV (extendable to 1TeV). A schematic view of the accelerator complex, indicating the location of the major sub-systems [26, 27, 28, 29, 30] is shown on Figure 2. and a more detailed layout on Figure 3.

One sees on the schematic layout

- a polarised electron source based on a photocathode DC gun.
- a polarised positron source in which positrons are obtained from electron-positron pairs by converting high-energy photons produced by passing the high-energy main electron beam through an undulator.
- 5 GeV electron and positron damping rings (DR) with a circumference of 3.2 km, housed in a common tunnel.
- beam transport from the damping rings to the main linacs, followed by a two-stage bunchcompressor system prior to injection into the main linac.
- two 11 km main linacs, utilising 1.3 GHz SCRF cavities operating at an average gradient of 31.5 MV/m, with a pulse length of 1.6 ms
- two beam-delivery systems, each 2.2 km long, which bring the beams into collision with a 14 mrad crossing angle, at a single interaction point which can be occupied by two detectors in a so-called push-pull configuration

A summary table of the parameters can be found on Figure 4. One should emphasize the synergy between the ILC and the European X-ray Free Electron Laser (XFEL) [31] which is currently under construction at DESY and will begin operation soon.

2.3.2. The Compact Linear Collider (CLIC). An overview of the CLIC layout can be found on Figure 5.

CLIC [33, 34, 32, 35] is based on high gradient normal-conducting accelerating structures where the RF power for the acceleration of the colliding beams is

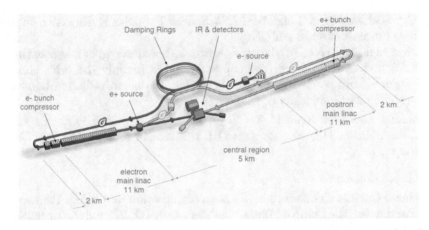

Figure 2. Schematic layout of the ILC, indicating all the major subsystems (not to scale) [26].

extracted from a high-current Drive Beam that runs parallel with the main linac. The focus of CLIC Research and Development over the last years has been on addressing a set of key feasibility issues that are essential for proving the fundamental validity of the CLIC concept. Several larger system tests have been performed to validate the two-beam scheme, and of particular importance are the results from the CLIC test facility at CERN (CTF3) which have demonstrated the two-beam acceleration at gradients exceeding 100 MV/m.The CLIC accelerator can be built in energy stages, as shown in Figure 6, re-using the existing equipment for each new stage. At each energy stage the centre of mass energy can be tuned to lower values within a range of approximately a factor three with limited loss in luminosity performance. The ultimate choice of the CLIC energy stages will be driven by the physics aims, where further input from LHC data, in particular 14 TeV data, is expected. The recent LHC Higgs discovery makes an initial energy stage around 350 GeV to 375 GeV very attractive, but final choices will depend on further LHC findings. CLIC main parameters can be found in Figure 7.

The yearly energy and power consumption is shown in Figure 8.

2.4. Future Circular Colliders

CERN is undertaking an integral design study for post-LHC particle accelerator options in a global context. The Future Circular Collider (FCC) [36] puts an emphasis on proton-proton and electron-positron (lepton) high-energy frontier machines.

This study is exploring the potential of hadron and lepton circular colliders, performing an in-depth analysis of infrastructure and operation concepts and considering the technology research and development programs that would be required to build a future circular collider. A conceptual design report will be delivered before the end of 2018, in time for the next update of the European Strategy for

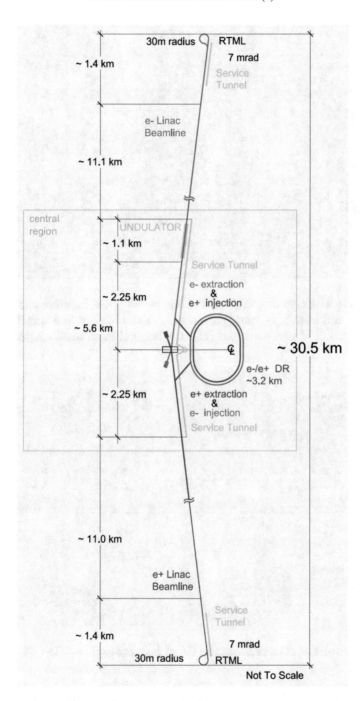

FIGURE 3. Layout of the ILC complex at $\sqrt{s} = 0.5$ TeV [29].

Centre-of-mass energy			Baseline 500 GeV Machine			1st Stage	L Upgrade	E_{CM} Upgrade	
								A	B
	E_{CM}	GeV	250	350	500	250	500	1000	1000
Collision rate	f_{rep}	Hz	5	5	5	5	5	4	4
Electron linac rate	f_{linac}	Hz	10	5	5	10	5	4	4
Number of bunches	n_b		1312	1312	1312	1312	2625	2450	2450
Bunch population	N	$\times 10^{10}$	2.0	2.0	2.0	2.0	2.0	1.74	1.74
Bunch separation	Δt_b	ns	554	554	554	554	366	366	366
Pulse current	I_{beam}	mA	5.8	5.8	5.8	5.8	8.8	7.6	7.6
Main linac average gradient	G_a	MV m^{-1}	14.7	21.4	31.5	31.5	31.5	38.2	39.2
Average total beam power	P_{beam}	MW	5.9	7.3	10.5	5.9	21.0	27.2	27.2
Estimated AC power	P_{AC}	MW	122	121	163	129	204	300	300
RMS bunch length	σ_z	mm	0.3	0.3	0.3	0.3	0.3	0.250	0.225
Electron RMS energy spread	$\Delta p/p$	%	0.190	0.158	0.124	0.190	0.124	0.083	0.085
Positron RMS energy spread	$\Delta p/p$	%	0.152	0.100	0.070	0.152	0.070	0.043	0.047
Electron polarisation	P_-	%	80	80	80	80	80	80	80
Positron polarisation	P_+	%	30	30	30	30	30	20	20
Horizontal emittance	$\gamma \epsilon_x$	μm	10	10	10	10	10	10	10
Vertical emittance	$\gamma \epsilon_y$	nm	35	35	35	35	35	30	30
IP horizontal beta function	β_x^*	mm	13.0	16.0	11.0	13.0	11.0	22.6	11.0
IP vertical beta function	β_y^*	mm	0.41	0.34	0.48	0.41	0.48	0.25	0.23
IP RMS horizontal beam size	σ_x^*	nm	729.0	683.5	474	729	474	481	335
IP RMS veritcal beam size	σ_y^*	nm	7.7	5.9	5.9	7.7	5.9	2.8	2.7
Luminosity	L	$\times 10^{34}$ cm^{-2}s^{-1}	0.75	1.0	1.8	0.75	3.6	3.6	4.9
Fraction of luminosity in top 1%	$L_{0.01}/L$		87.1%	77.4%	58.3%	87.1%	58.3%	59.2%	44.5%
Average energy loss	δ_{BS}		0.97%	1.9%	4.5%	0.97%	4.5%	5.6%	10.5%
Number of pairs per bunch crossing	N_{pairs}	$\times 10^3$	62.4	93.6	139.0	62.4	139.0	200.5	382.6
Total pair energy per bunch crossing	E_{pairs}	TeV	46.5	115.0	344.1	46.5	344.1	1338.0	3441.0

FIGURE 4. Summary table of the 250 ∼ 500 GeV baseline and luminosity and energy upgrade parameters. Also included is a possible 1st stage 250 GeV parameter set (half the original main linac length) [26].

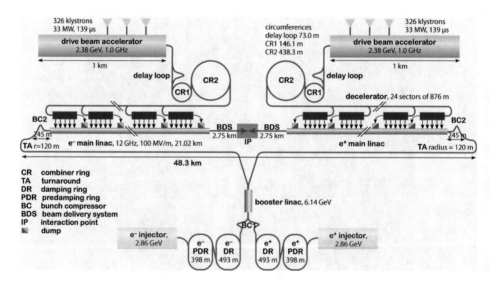

FIGURE 5. Overview of the CLIC layout at $\sqrt{s} = 3$ TeV [32].

Particle Physics. Here we will not discuss FCC-he, High-Energy LHC (HL-LHC) [21] or LHeC [37, 38].

Parameter	Symbol	Unit	Stage 1	Stage 2	Stage 3
Center-of-mass energy	$\sqrt{s}$	GeV	350	1400	3000
Integrated luminosity	$\mathscr{L}_{int}$	ab^{-1}	0.5	1.5	2.0

FIGURE 6. Center of mass energy and assumed integrated luminosity for the different CLIC machine stages [35]. The integrated luminosities correspond each to four or five years of operation of a fully commissioned machine running 200 days per year with an effective up-time of 50%.

Description [units]	500 GeV	3 TeV
Total (peak 1%) luminosity	$2.3\,(1.4)\times10^{34}$	$5.9\,(2.0)\times10^{34}$
Total site length [km]	13.0	48.4
Loaded accel. gradient [MV/m]	80	100
Main Linac RF frequency [GHz]	12	
Beam power/beam [MW]	4.9	14
Bunch charge [10^9 e$^+$/e$^-$]	6.8	3.72
Bunch separation [ns]	0.5	
Bunch length [μm]	72	44
Beam pulse duration [ns]	177	156
Repetition rate [Hz]	50	
Hor./vert. norm. emitt. [$10^{-6}/10^{-9}$m]	2.4/25	0.66/20
Hor./vert. IP beam size [nm]	202/2.3	40/1
Beamstrahlung photons/electron	1.3	2.2
Hadronic events/crossing at IP	0.3	3.2
Coherent pairs at IP	200	6.8×10^8

FIGURE 7. CLIC main parameters for centre of mass energies of 500 GeV and 3 TeV [33].

	Power [MW]	Days	Energy [TWh]
Nominal operation mode	582	177	2.47
Fault-induced downtime	60	44	0.06
Programmed stops	60	144	0.21
Energy consumption per year			2.74

FIGURE 8. Yearly energy and power consumption for the nominal 3 TeV CLIC [33].

2.4.1. FCC-hh. Circular proton-proton colliders are the main, and possibly only, tool available for exploring particle physics in the energy range of tens of TeV.

The bending radius ρ of a relativistic particle of charge e and momentum p is related to the magnetic field of strength B by $p = eB\rho$. Therefore the energy of pp collisions can be raised by increasing the strength of the dipole magnets, or the bending radius ρ, and, thereby, the ring circumference. The proton-proton FCC collider (FCC-hh) design combines both approaches. Specifically, the FCC ring circumference of about 100 km would enable pp collisions of 50 TeV in the centre of mass with the present 8.3 T LHC magnets (made with $NbTi$ superconductor), of 100 TeV with 16 T magnets and of 125 TeV with 20 T magnets. The main technological challenge is therefore magnets: one should replace the magnets. Nb_3Sn superconductor can reach a proactical magnetic field up to 15 T and, as discussed above, few Nb_3Sn magnets are planned for HL-LHC, which will represent an important milestone towards the FCC. High Temperature Superconductor (HTS) materials, like yttrium copper oxyde YBCO could be used in an hybrid coil design in order to get up to 20 T. A sketch of such a magnet is shown on Figure 9. One should note that there are very important technological challenges here, as

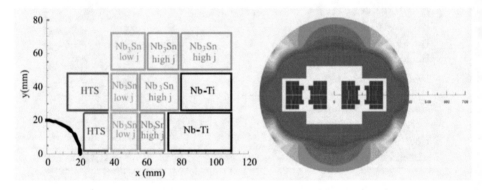

FIGURE 9. Sketch of cross section of a 20 T magnet with hybrid coil [21].

shown on Figure 10 where the progress of accelerator magnets for hadron colliders is shown.

2.4.2. FCC-ee. The discovery of a BEH boson at an energy reachable by a collider slightly more energetic than LEP2, together with the excellent performance of the two B factories PEP-II and KEKB, have led to new proposals [39, 40, 41] for a next generation e^+e^- collider. In order to serve as a BEH factory such a collider needs to be able to operate at at least a centre of mass energy of 240 GeV (for efficient $e^-e^- \rightarrow ZH$ production), i.e., 15% above the LEP2 peak energy. Reaching even higher energies,e.g., up to 350 GeV centre of mass, for $t\bar{t}$ production, or even maybe 500 GeV for Zhh and $t\bar{t}h$ studies, might be possible for a new ring of larger circumference. The preliminary power consumption is estimated to be about 300 MW at 350 GeV centre of mass [40, 15].

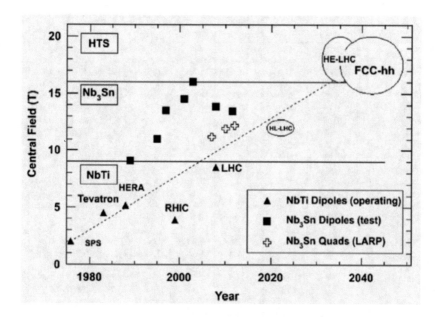

FIGURE 10. Progress of accelerator magnets for hadron colliders [23].

2.4.3. CepC. China is currently thinking to an e^+e^- collider, where the current design, CepC [42, 43],is a single ring collider two times larger than LEP. It can be associated with a 30–50 TeV proton collider, called SppC.

2.4.4. Comparison of various proposals. One can find in Figure 11 the comparison of various parameters.

Projected e^+e^- luminosities can be found in Figure 12.

3. New physics in the scalar sector

Only new physics in the scalar sector will be described rapidly here. Discussions of new physics outside the scalar sector can be found in reports of each future facilities or elsewhere (for HL-LHC see [44, 45]). One can look at reviews of the BEH boson inside [46] and outside [47, 48, 49] the Standard Model. It is outside the scope of this article to give details. I just remind that the most economical low-energy supersymmetric extension of the SM is the Minimal Supersymmetric Standard Model (MSSM) which is itself one of the 2HDM, 2 (BE)Higgs Doublet Models. Five physical bosons are left in the spectrum, one charged pair $H^{\pm}$, one CP-odd scalar, A, and two CP-even states, H and h. There are two free parameters at the tree level, which can be taken as M_A and $\tan\beta$. Deviations from the SM are often parametrized as scale factors (κ) of BEH couplings relative to their SM values [50].

parameter	LHC (pp) design	FCC-hh	LEP2 achieved	FCC-ee (TLEP)					CepC
				Z	Z (cr. w.)	W	H	$t\bar{t}$	
species	pp	pp	e^+e^-	e^+e^-	e^+e^-	e^+e^-	e^+e^-	e^+e^-	e^+e^-
E_{beam} [GeV]	7,000	50,000	104	45.5	45	80	120	175	120
circumf. [km]	26.7	100	26.7	100	100	100	100	100	54
current [mA]	584	500	3.0	1450	1431	152	30	6.6	16.6
no. of bunches, n_b	2808	10600	4	16700	29791	4490	1360	98	50
N_b [10^{11}]	1.15	1.0	4.2	1.8	1.0	0.7	0.46	1.4	3.7
ϵ_x [nm]	0.5	0.04	22	29	0.14	3.3	0.94	2	6.8
ϵ_y [pm]	500	41	250	60	1	7	2	2	20
β_x^* [m]	0.55	1.1	1.2	0.5	0.5	0.5	0.5	1.0	0.8
β_y^* [mm]	550	1100	50	1	1	1	1	1	1.2
σ_x^* [μm]	16.7	6.8	162	121	8	26	22	45	74
σ_y^* [μm]	16.7	6.8	3.5	0.25	0.032	0.13	0.044	0.045	0.16
θ_c [mrad]	0.285	0.074	0	0	30	0	0	0	0
f_{rf} [MHz]	400	400	352	800	300	800	800	800	700
V_{rf} [GV]	0.016	>0.020	3.5	2.5	0.54	4	5.5	11	6.87
α_c [10^{-5}]	32	11	14	18	2	2	0.5	0.5	4.15
$\delta_{\mathrm{rms}}^{\mathrm{SR}}$ [%]	—	—	0.16	0.04	0.04	0.07	0.10	0.14	0.13
$\sigma_{z,\mathrm{rms}}^{\mathrm{SR}}$ [mm]	—	—	11.5	1.64	1.9	1.01	0.81	1.16	2.3
$\delta_{\mathrm{rms}}^{\mathrm{tot}}$ [%]	0.003	0.004	0.16	0.06	0.12	0.09	0.14	0.19	0.16
$\sigma_{z,\mathrm{rms}}^{\mathrm{tot}}$ [mm]	75.5	80	11.5	2.56	6.4	1.49	1.17	1.49	2.7
F_{hg}	1.0	1.0	0.99	0.64	0.94	0.79	0.80	0.73	0.61
$\tau_{\parallel}$ [turns]	10^9	10^7	31	1320	1338	243	72	23	40
ξ_x/IP	0.0033	0.005	0.04	0.031	0.032	0.060	0.093	0.092	0.103
ξ_y/IP	0.0033	0.005	0.06	0.030	0.175	0.059	0.093	0.092	0.074
no. of IPs, n_{IP}	3 (4)	2 (4)	4	4	4	4	4	4	2
L/IP [10^{34}/cm^2/s]	1	5	0.01	28	219	12	6	1.7	1.8
τ_{beam} [min]	2760	1146	300	287	38	72	30	23	57
P_{SR}/beam [MW]	0.0036	2.4	11	50	50	50	50	50	50
energy / beam [MJ]	392	8400	0.03	22	22	4	1	0.4	0.3

FIGURE 11. Parameters of the proposed FCC-hh, FCC-ee/TLEP and CepC, compared with LEP2 and the LHC Design [15].

In general the size of couplings modifications from the SM values is of the order of few % when new physics is at a scale of 1 TeV [51]. and decreases when the new physics scale increases. This set the scale for the uncertainties that one aims.

Tests of new physics in the BEH sector can be done, either by studying the decays of the already discovered boson, or looking at new bosons.

Detectors will not be described here, but information can be found in the reports associated to the facilities.

3.1. Next runs of the LHC

One can find more information in recent workshops [24] or reviews [52, 53]. One sees in Figure 13 and 14 the expected precisions from CMS [45] on the measurements of the signal strengths (ratio of the cross section divided by the SM predicted cross section) and on the coupling scale factors κ. Extrapolations are presented under two uncertainty scenarios. In Scenario 1, all systematic uncertainties are left unchanged. In Scenario 2, the theoretical uncertainties are scaled by a factor

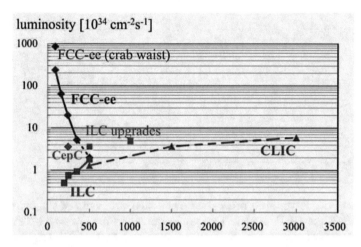

FIGURE 12. Projected total electron-positron luminosities vs. c.m. energy for various proposed colliders [15]. The luminosities of FCC-ee and CepC are summed respectively over 4 or 2 IPs.

of 1/2, while other systematic uncertainties are scaled by the square root of the integrated luminosity. The comparison of the two uncertainty scenarios indicates a range of possible future measurements.

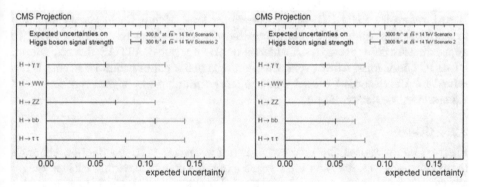

FIGURE 13. Estimated precision on the measurements of the signal strength for a SM-like BEH boson. The projections assume $\sqrt{s} = 14\ TeV$ and an integrated dataset of 300 fb^{-1} (left) and 3000 fb^{-1} (right). The projections are obtained with the two uncertainty scenarios described in the text [45].

Figure 15, from ATLAS, [54] shows the relative uncertainty on the signal strenth μ defined as the ratio of the measured cross section by the SM cross section for various BEH decay final states. The uncertainties are quite good.

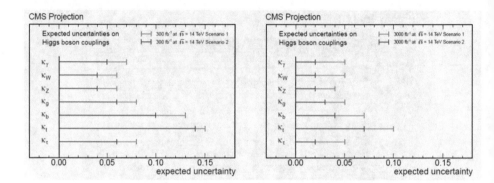

FIGURE 14. Estimated precision on the measurements of the couplings for a SM-like BEH boson. The projections assume $\sqrt{s} = 14$ TeV and an integrated dataset of 300 fb^{-1} (left) and 3000 fb^{-1} (right). The projections are obtained with the two uncertainty scenarios described in the text [45].

Some tests on additional bosons of the BEH sector can also be done, looking at $A/H \to \tau\tau$, $A/H \to t\bar{t}$ $A \to Zh$, $H \to hh$, $H \to ZZ$.

Studies are presented on the prospects for the observation of Higgs pair production in the channel $H \to \gamma\gamma$ $H \to b\bar{b}$ using an upgraded ATLAS detector, assuming a dataset comprising 3000 fb^{-1} of 14 TeV proton-proton collisions at the High-Luminosity LHC (HL-LHC). Generator-level Monte Carlo events are used to perform this study, with parameterised efficiencies and resolution applied to approximate the expected performance of the upgraded ATLAS detector under HL-LHC conditions. After event selection, a signal yield of around 8 events is obtained for the Standard Model scenario, corresponding to a signal significance of 1.3 standard deviation [55].

3.2. FCC-hh

Figure 16 summarises the increase in rate for several BEH production channels in pp collisions, as a function of the centre of mass energy, covering various ranges of possibilities that are curently discussed. Final states with the largest invariant mass (like tth and hh) benefit the most from the energy increase.

3.3. e^+e^- colliders

The cross sections as a function of the centre of mass energy are shown in Figure 17.

A key production mode is $e^+e^- \to Zh$ where events can be detected inclusively, completely independent of the BEH decay mode by tagging the Z via $Z \to e^+e^-$ or $\mu^+\mu^-$ and requiring that the recoil mass is consistent with the BEH mass. The normalisation of this rate allows a precise measurement of the coupling

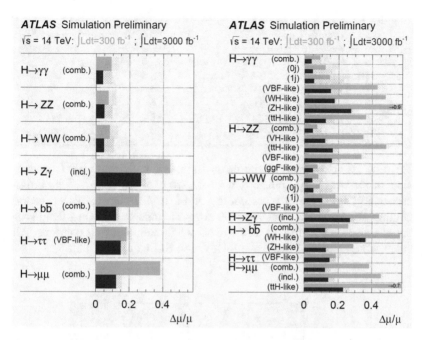

FIGURE 15. Relative uncertainty on the signal strength μ for all Higgs final states considered in this note in the different experimental categories used in the combination, assuming a SM BEH boson with a mass of 125 GeV expected with 300 fb^{-1} and 3000 fb^{-1} 14 TeV LHC data. The uncertainty pertains to the number of events passing the experimental selection, not to the particular BEH boson process targeted. The hashed areas indicate the increase of the estimated error due to current theory systematic uncertainties. The abbreviation (comb.) indicates that the precision on μ is obtained from the combination of the measurements from the different experimental sub-categories for the same final state, while (incl.) indicates that the measurement from the inclusive analysis was used. The left side shows only the combined signal strength in the considered final states, while the right side also shows the signal strength in the main experimental sub-categories within each final state [54].

of the Z to the BEH boson. One can then obtain *absolute* measurements of *all* possible branching ratios and the BEH total width.

The various couplings of the BEH boson to different particles can be obtained in e^+e^- colliders with very good precisions. For instance Figure 18 shows the results of a *full* ILC program.

Expected accuracies are also shown in Figure 19 and are very good.

Process	σ (14 TeV)	R (33)	R (40)	R (60)	R (80)	R (100)
gg→H	50.4 pb	3.5	4.6	7.8	11	15
qq→qqH	4.40 pb	3.8	5.2	9.3	14	19
q$\bar{q}$→WH	1.63 pb	2.9	3.6	5.7	7.7	10
q$\bar{q}$→ZH	0.90 pb	3.3	4.2	6.8	10	13
pp→HH	33.8 fb	6.1	8.8	18	29	42
pp→ttH	0.62 pb	7.3	11	24	41	61

FIGURE 16. Evolution of the cross-sections for different BEH production processes in pp collisions with centre of mass energy. The cross-sections at a centre of mass energy of 14 TeV is shown in the second column,and the ratios between the cross sections at the considered centre of mass energy and 14 TeV are shown in the following columns. All rates assume a BEH mass of 125GeV and SM couplings [23].

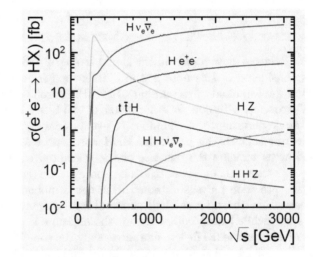

FIGURE 17. The center-of-mass dependencies of the cross sections for the main Higgs production processes at an e^+e^- collider. The values shown correspond to unpolarized beams and do not include the effects of initial-stare radiation (ISR) or beamstrahlung [35].

3.4. Comparisons

Expected predictions on the BEH couplings are shown in Figure 20 and match the required uncertainties.

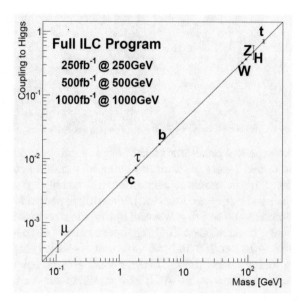

FIGURE 18. Expected precision from the full ILC program of tests of the Standard Model prediction that the BEH coupling to each particle is proportional to its mass [27].

Mode	LHC	ILC(250)	ILC500	ILC(1000)
WW	4.1 %	1.9 %	0.24 %	0.17 %
ZZ	4.5 %	0.44 %	0.30 %	0.27 %
$b\bar{b}$	13.6 %	2.7 %	0.94 %	0.69 %
gg	8.9 %	4.0 %	2.0 %	1.4 %
$\gamma\gamma$	7.8 %	4.9 %	4.3 %	3.3 %
$\tau^+\tau^-$	11.4 %	3.3 %	1.9 %	1.4 %
$c\bar{c}$	–	4.7 %	2.5 %	2.1 %
$t\bar{t}$	15.6 %	14.2 %	9.3 %	3.7 %
$\mu^+\mu^-$	–	–	–	16 %
self	–	–	104%	26 %
BR(invis.)	< 9%	< 0.44 %	< 0.30 %	< 0.26 %
$\Gamma_T(h)$	20.3%	4.8 %	1.6 %	1.2 %

FIGURE 19. Expected accuracies for Higgs boson couplings. For the invisible branching ratio, the numbers quoted are 95% confidence upper limits. The four columns refer to: LHC, 300 fb^{-1}, 1 detector; ILC at 250 GeV, with 250 fb^{-1}; ILC at 500 GeV, with 500 fb^{-1}; ILC at 1000 GeV, with 1000 fb^{-1}. Each column includes the stated data set and all previous ones [27]

Facility	LHC	HL-LHC	ILC500	ILC500-up	ILC1000	ILC1000-up	CLIC	TLEP (4 IPs)
$\sqrt{s}$ (GeV)	14,000	14,000	250/500	250/500	250/500/1000	250/500/1000	350/1400/3000	240/350
$\int \mathcal{L}dt$ (fb^{-1})	300/expt	3000/expt	250+500	1150+1600	250+500+1000	1150+1600+2500	500+1500+2000	10,000+2600
κ_γ	5 − 7%	2 − 5%	8.3%	4.4%	3.8%	2.3%	−/5.5/<5.5%	1.45%
κ_g	6 − 8%	3 − 5%	2.0%	1.1%	1.1%	0.67%	3.6/0.79/0.56%	0.79%
κ_W	4 − 6%	2 − 5%	0.39%	0.21%	0.21%	0.2%	1.5/0.15/0.11%	0.10%
κ_Z	4 − 6%	2 − 4%	0.49%	0.24%	0.50%	0.3%	0.49/0.33/0.24%	0.05%
κ_ℓ	6 − 8%	2 − 5%	1.9%	0.98%	1.3%	0.72%	3.5/1.4/<1.3%	0.51%
$\kappa_d = \kappa_b$	10 − 13%	4 − 7%	0.93%	0.60%	0.51%	0.4%	1.7/0.32/0.19%	0.39%
$\kappa_u = \kappa_t$	14 − 15%	7 − 10%	2.5%	1.3%	1.3%	0.9%	3.1/1.0/0.7%	0.69%

FIGURE 20. Expected precisions on the Higgs couplings and total width from a constrained 7-parameter fit assuming no non-SM production or decay modes. The fit assumes generation universality (the couplings for u,c,t (d,s,b and e,μ,τ) are identical). The ranges shown for LHC and HL-LHC represent the conservative and optimistic scenarios for systematic and theory uncertainties. ILC numbers assume $(e-; e+)$ polarizations of $(-0.8; -0.3)$ at 250 and 500 GeV and $(-0.8; 0.2)$ at 1000 GeV, plus a 0.5% theory uncertainty. CLIC numbers assume polarizations of $(-0.8; 0)$ for energies above 1 TeV. TLEP numbers assume unpolarized beams [51]

4. Conclusions

One hope to have interesting results in the next runs of the LHC (runs 2 and 3, and, after 2025, HL-LHC).

The longer term future should also provide major tests of physics outside the Standard Model in the scalar sector, with e^+e^- (circular or linear) colliders or pp colliders.

Acknowledgement

I thank D. Delgove for his help. I thank also G. Arduini, P. Bambade, A. Blondel, O. Bruning, J.-B. De Vivie, A. Djouadi, A. Falkowski, C. Grojean, I. Laktineh, L. Linssen, L. Mapelli, L. Nisati, D. Schulte, J. Wenninger and F. Zimmermann for discussions.

References

[1] G. Aad *et al.* [ATLAS Collaboration], "Observation of a new particle in the search for the Standard Model Higgs boson with the ATLAS detector at the LHC," Phys. Lett. B **716**, 1 (2012) [arXiv:1207.7214 [hep-ex]].

[2] S. Chatrchyan *et al.* [CMS Collaboration], "Observation of a new boson at a mass of 125 GeV with the CMS experiment at the LHC," Phys. Lett. B **716**, 30 (2012) [arXiv:1207.7235 [hep-ex]].

[3] O. Bruning, H. Burkhardt and S. Myers, "The large hadron collider," Prog. Part. Nucl. Phys. **67**, 705 (2012).

[4] L. Evans and P. Bryant, "LHC Machine," JINST **3**, S08001 (2008).

[5] F. Englert, "The BEH mechanism and its scalar boson," Annalen Phys. **526**, 201 (2014).

[6] P.W. Higgs, "Evading the Goldstone theorem," Annalen Phys. **526**, 211 (2014).

[7] [Royal Swedish Academy of Sciences Collaboration], "The BEH-mechanism, interactions with short range forces and scalar particles," AAPPS Bulletin **23**, 6, 3 (2013).

[8] S. Weinberg, "The Making of the standard model," Eur. Phys. J. C **34**, 5 (2004), [hep-ph/0401010].

[9] N. Andari, "Observation of a BEH-like boson decaying into two photons with the ATLAS detector at the LHC," CERN-THESIS-2012-144, LAL12-300.

[10] N. Craig, "The State of Supersymmetry after Run I of the LHC," arXiv:1309.0528 [hep-ph].

[11] Y. Alexahin, C.M. Ankenbrandt, D.B. Cline, A. Conway, M.A. Cummings, V. Di Benedetto, E. Eichten and C. Gatto *et al.*, "Muon Collider Higgs Factory for Snowmass 2013," arXiv:1308.2143 [hep-ph].

[12] V.I. Telnov, "Comments on photon colliders for Snowmass 2013," arXiv:1308.4868 [physics.acc-ph].

[13] S.M. Hooker, "Developments in laser-driven plasma accelerators," Nature Photon. **7**, 775 (2013), [arXiv:1406.5118 [physics.plasm-ph]].

[14] A. Blondel, A. Chao, W. Chou, J. Gao, D. Schulte and K. Yokoya, "Report of the ICFA Beam Dynamics Workshop 'Accelerators for a Higgs Factory: Linear vs. Circular' (HF2012)," arXiv:1302.3318 [physics.acc-ph].

[15] F. Zimmermann, M. Benedikt, D. Schulte and J. Wenninger, "Challenges for Highest Energy Circular Colliders," IPAC-2014-MOXAA01.

[16] http://council.web.cern.ch/council/en/EuropeanStrategy/ ESParticlePhysics.html.

[17] http://www.usparticlephysics.org/p5/.

[18] http://higgshunting.fr.

[19] 11th ICFA Seminar in Beijing (october 2014), http://indico.ihep.ac.cn/conferenceOtherViews.py?view=standard &confId=3867.

[20] S. Myers, "Summary of the RLIUP Workshop, Archamps 29th–31st October 2013", https://indico.cern.ch/event/281478/overview.

[21] E. Todesco, M. Lamont and L. Rossi, "High luminosity LHC and high energy LHC," Proceedings, CMS Workshop: Perspectives on Physics and on CMS at Very High Luminosity, HL-LHC: Alushta, Crimea, Ukraine, May 2831, 2012.

[22] G. Arduini, "LHC Machine Upgrade," IFD2014, INFN Workshop on Future Detectors for HL-LHC, Trento, http://events.unitn.it/en/ifd2014.

[23] W. Barletta, M. Battaglia, M. Klute, M. Mangano, S. Prestemon, L. Rossi and P. Skands, "Future hadron colliders: From physics perspectives to technology Research and Development," Nucl. Instrum. Meth. A **764**, 352 (2014).

[24] ECFA High Luminosity LHC Experiments Workshop – 2014 (october 2014), https://indico.cern.ch/event/315626/other-view?view=standard.

[25] https://www.linearcollider.org.

[26] T. Behnke, J. E. Brau, B. Foster, J. Fuster, M. Harrison, J. M. Paterson, M. Peskin and M. Stanitzki *et al.*, "The International Linear Collider Technical Design Report – Volume 1: Executive Summary," arXiv:1306.6327 [physics.acc-ph].

[27] H. Baer, T. Barklow, K. Fujii, Y. Gao, A. Hoang, S. Kanemura, J. List and H. E. Logan *et al.*, "The International Linear Collider Technical Design Report – Volume 2: Physics," arXiv:1306.6352 [hep-ph].

[28] C. Adolphsen, M. Barone, B. Barish, K. Buesser, P. Burrows, J. Carwardine, J. Clark and H. M. Durand *et al.*, "The International Linear Collider Technical Design Report – Volume 3.I: Accelerator & in the Technical Design Phase," arXiv:1306.6353 [physics.acc-ph].

[29] C. Adolphsen, M. Barone, B. Barish, K. Buesser, P. Burrows, J. Carwardine, J. Clark and H.M. Durand *et al.*, "The International Linear Collider Technical Design Report – Volume 3.II: Accelerator Baseline Design," arXiv:1306.6328 [physics.acc-ph].

[30] T. Behnke, J.E. Brau, P.N. Burrows, J. Fuster, M. Peskin, M. Stanitzki, Y. Sugimoto and S. Yamada *et al.*, "The International Linear Collider Technical Design Report – Volume 4: Detectors," arXiv:1306.6329 [physics.ins-det]. INSPIRE as of 02 Nov 2014.

[31] M. Altarelli, R. Brinkmann, M. Chergui, W. Decking, B. Dobson, S. Dusterer, G. Grubel and W. Graeff *et al.*, "XFEL: The European X-Ray Free-Electron Laser. Technical design report," DESY-06-097.

[32] P. Lebrun, L. Linssen, A. Lucaci-Timoce, D. Schulte, F. Simon, S. Stapnes, N. Toge and H. Weerts *et al.*, "The CLIC Programme: Towards a Staged e^+e^- Linear Collider Exploring the Terascale: CLIC Conceptual Design Report," arXiv:1209.2543 [physics.ins-det].

[33] M. Aicheler, P. Burrows, M. Draper, T. Garvey, P. Lebrun, K. Peach and N. Phinney *et al.*, "A Multi-TeV Linear Collider Based on CLIC Technology: CLIC Conceptual Design Report," CERN-2012-007, SLAC-R-985, KEK-Report-2012-1, PSI-12-01, JAI-2012-001.

[34] L. Linssen, A. Miyamoto, M. Stanitzki and H. Weerts, "Physics and Detectors at CLIC: CLIC Conceptual Design Report," arXiv:1202.5940 [physics.ins-det], Nov 2014.

[35] H. Abramowicz *et al.* [CLIC Detector and Physics Study Collaboration], "Physics at the CLIC e^+e^- Linear Collider – Input to the Snowmass process 2013," arXiv:1307.5288 [hep-ex].

[36] http://cern.ch/fcc.

[37] J.L. Abelleira Fernandez, C. Adolphsen, P. Adzic, A.N. Akay, H. Aksakal, J.L. Albacete, B. Allanach and S. Alekhin *et al.*, "A Large Hadron Electron Collider at CERN," arXiv:1211.4831 [hep-ex].

[38] J.L. Abelleira Fernandez *et al.* [LHeC Study Group Collaboration], "A Large Hadron Electron Collider at CERN: Report on the Physics and Design Concepts for Machine and Detector," J. Phys. G **39**, 075001 (2012), [arXiv:1206.2913 [physics.acc-ph]].

[39] A. Blondel and F. Zimmermann, "A High Luminosity e^+e^- Collider in the LHC tunnel to study the Higgs Boson," arXiv:1112.2518.

[40] M. Koratzinos, A.P. Blondel, R. Aleksan, O. Brunner, A. Butterworth, P. Janot, E. Jensen and J. Osborne *et al.*, "TLEP: A High-Performance Circular e^+e^- Collider to Study the Higgs Boson," arXiv:1305.6498 [physics.acc-ph].

[41] M. Bicer *et al.* [TLEP Design Study Working Group Collaboration], "First Look at the Physics Case of TLEP," JHEP **1401**, 164 (2014), [arXiv:1308.6176 [hep-ex]].

[42] http://cepc.ihep.ac.cn.

[43] X.C. Lou, " The CEPC-SppC Study Group in China, CFHEP Kick-off Meeting, Beijing", http://beijingcenterfuturecollider.wikispaces.com/events.

[44] [ATLAS Collaboration], "Search for Supersymmetry at the high luminosity LHC with the ATLAS experiment ATL-PHYS-PUB-2014-010".

[45] [CMS Collaboration], "Projected Performance of an Upgraded CMS Detector at the LHC and HL-LHC: Contribution to the Snowmass Process," arXiv:1307.7135.

[46] A. Djouadi, "The Anatomy of electro-weak symmetry breaking. I: The Higgs boson in the standard model," Phys. Rep. **457**, 1 (2008), [hep-ph/0503172].

[47] A. Djouadi, "The Anatomy of electro-weak symmetry breaking. II. The Higgs bosons in the minimal supersymmetric model," Phys. Rept. **459**, 1 (2008), [hep-ph/0503173].

[48] D. Rainwater, "Searching for the Higgs boson," hep-ph/0702124.

[49] G.C. Branco, P.M. Ferreira, L. Lavoura, M.N. Rebelo, M. Sher and J.P. Silva, "Theory and phenomenology of two-Higgs-doublet models," Phys. Rep. **516**, 1 (2012), [arXiv:1106.0034 [hep-ph]].

[50] A. David *et al.* [LHC Higgs Cross Section Working Group Collaboration], "LHC HXSWG interim recommendations to explore the coupling structure of a Higgs-like particle," arXiv:1209.0040 [hep-ph].

[51] S. Dawson, A. Gritsan, H. Logan, J. Qian, C. Tully, R. Van Kooten, A. Ajaib and A. Anastassov *et al.*, "Working Group Report: Higgs Boson," arXiv:1310.8361 [hep-ex].

[52] A. G. Holzner [for the ATLAS and CMS Collaborations], "Beyond standard model Higgs physics: prospects for the High Luminosity LHC," arXiv:1411.0322 [hep-ex].

[53] M. Vidal [CMS Collaboration], "Future prospects of Higgs Physics at CMS," arXiv:1409.1711 [hep-ex].

[54] [ATLAS Collaboration], "Projections for measurements of Higgs boson signal strengths and coupling parameters with the ATLAS detector at a HL-LHC, ATL-PHYS-PUB-2014-016 ".

[55] [ATLAS Collaboration], "Prospects for measuring Higgs pair production in the channel $H \to \gamma\gamma \; H \to b\bar{b}$ using the ATLAS detector at the HL-LHC, ATL-PHYS-PUB-2014-019 ".

Louis Fayard
LAL – Bât. 200
Université Paris-Sud
F-91898 Orsay cedex, France
e-mail: lfayard@in2p3.fr

The H Boson, 85–133
© 2017 Springer Basel AG

| Poincaré Seminar 2014

Implications of the H Boson Discovery

Abdelhak Djouadi

Abstract. The implications of the discovery of a scalar boson at the LHC with a mass of approximately 125 GeV are summarised in the context of the Standard Model of particle physics and its minimal supersymmetric extension, the MSSM. Discussed are the implications from the measured mass and production/decay rates of the observed particle and from the constraints in the search for the heavier scalar states at the LHC in the MSSM.

1. Introduction

The ATLAS and CMS historical discovery of a particle with a mass of approximately 125 GeV [1, 2] and properties that are compatible with those of a scalar boson [3, 4, 5, 6] has far reaching consequences not only for the Standard Model (SM) of the electroweak and strong interactions, but also for new physics models beyond it.

In the SM, electroweak symmetry breaking (EWSB) is achieved spontaneously via the Brout–Englert–Higgs mechanism [3], wherein the neutral component of an isodoublet scalar field acquires a non-zero vacuum expectation value. This gives rise to nonzero masses for the fermions and the electroweak gauge bosons, which are otherwise not allowed by the $SU(2) \times U(1)$ gauge symmetry. In the sector of the theory with broken symmetry, one of the four degrees of freedom of the original isodoublet field, corresponds to a physical particle: a scalar boson with $J^{PC} = 0^{++}$ quantum numbers under parity and charge conjugation. We will "sacrify to the tradition" and call this particle the H or Higgs boson in the following.

The couplings of the Higgs boson to the fermions and gauge bosons are related to the masses of these particles and are thus decided by the symmetry breaking mechanism. In contrast, the mass of the Higgs boson itself M_H, although expected to be in the vicinity of the EWSB scale $v \approx 250$ GeV, is undetermined. Let us summarise the known information on this parameter before the start of the LHC.

A direct information was the lower limit $M_H \gtrsim 114$ GeV at 95% confidence level (CL) established at LEP2 [7, 8]. Furthermore, the high accuracy of the electroweak data measured at LEP, SLC and the Tevatron [7] provides an indirect sensitivity to M_H: the Higgs boson contributes logarithmically, $\propto \log(M_H/M_W)$, to the radiative corrections to the W and Z boson propagators. A global fit of the electroweak precision data yields the value $M_H = 92^{+34}_{-26}$ GeV, corresponding to a 95% CL upper limit of $M_H \lesssim 161$ GeV [9]. Another analysis, using a different fitting program gives a comparable value $M_H = 96^{+31}_{-24}$ GeV [10]. In both cases, the Higgs mass given above are when the limits from direct searches are not included in the global fits.

From the theoretical side, the presence of this new weakly coupled degree of freedom is a crucial ingredient for a unitary electroweak theory. Indeed, the SM without the Higgs particle is not self-consistent at high energies as it leads to scattering amplitudes of the massive electroweak gauge bosons that grow with the square of the center of mass energy and perturbative unitarity would be lost at energies above the TeV scale. In fact, even in the presence of a Higgs boson, the W and Z bosons could interact very strongly with each other and, imposing the unitarity requirement in the W and Z boson high-energy scattering amplitudes leads to the important Higgs mass bound $M_H \lesssim 700$ GeV [11], implying that the particle is kinematically accessible at the LHC. It is interesting to note, as an aside, that just the requirement of perturbative unitarity in these scattering amplitudes leads to a model with exactly the same particle content and couplings as the SM [12].

Another theoretical constraint emerges from the fact that the self-coupling of the Higgs boson, which is proportional to M_H^2, evolves with energy by virtue of quantum fluctuations (virtual fermions, gauge and Higgs bosons are exchanged in the coupling among three or four Higgs particles). This evolution is rather strong and at some stage, the coupling becomes very large and even infinite and the theory completely looses its predictability. If the energy scale up to which the couplings remains finite and the SM effectively valid, is of the order of the Higgs mass itself, M_H should be approximately $M_H \lesssim 650$ GeV when only the first terms of the perturbation series are included [13]. This value is remarkably close to the one obtained from numerical simulations in lattice gauge theory where the theory can be solved exactly. On the other hand, for small values of the self-coupling, and hence of the Higgs boson mass, the quantum fluctuations tend to drive the coupling to negative values and, thus, completely destabilize the scalar Higgs potential to the point where the minimum is not stable anymore. Requiring that the self-coupling stays positive and the minimum stable up to energies of about 1 TeV implies that the Higgs boson should have a mass above approximately 70 GeV. However, if the SM is to be extended to ultimate scales, such as for instance the Planck scale $M_P \sim 10^{18}$ GeV, these requirements on the self-coupling from finiteness and positivity become much more constraining and the Higgs mass should lie in the range 130 GeV $\lesssim M_H \lesssim$ 180 GeV [13, 14]. This is a rather narrow margin that is close to the one obtained from the direct and indirect experimental constraints.

The discovery of the Higgs particle with a mass of 125 GeV, a value that makes the SM perturbative, unitary and extrapolable to the highest possible scales, is therefore a consecration of the model and crowns its past success in describing all experimental data available. Nevertheless, the SM is far from being considered to be perfect in many respects. Indeed, it does not explain the proliferation of fermions and the large hierarchy in their mass spectra and, in particular, it does not say much about the observed small neutrino masses. The SM does not unify in a satisfactory way the electromagnetic, weak and strong forces, as one has three different symmetry groups with three different coupling constants which, in addition, shortly fail to meet at a common value during their evolution with the energy scale; it also ignores the fourth force, the gravitational interaction. Furthermore, it does not contain a massive, electrically neutral, weakly interacting and absolutely stable particle which would account for dark matter which is expected to represent 25% of the energy content of the Universe and fails to explain the baryon asymmetry in the Universe: why there are by far more particles than antiparticles.

However, the main problem that leads to the widespread belief that the SM is simply an effective theory, valid only at the energy scales that have been explored so far and should be replaced by a more fundamental theory at the TeV scale, is related to the particular status of the Higgs boson. Contrary to fermions and gauge bosons, the Higgs particle has a mass that cannot be protected against quantum corrections. These corrections are quadratic in the new physics which serves as a cut-off and hence, tend to drive the Higgs mass to very large values, ultimately to the Planck scale, while we need it to be close to the 100 GeV range. Thus, the SM cannot be extrapolated up to energies higher than the TeV scale where some new physics should emerge. This is the main reason which makes that something new, in addition to the Higgs particle, is expected to manifest itself at the LHC.

Among the many possibilities for this new physics beyond the SM[1], the option that emerges in the most natural way is Supersymmetry (SUSY) [15]. SUSY combines internal gauge with space-time symmetries and relates fermions and bosons: to each particle, it predicts the existence of a super-partner which should have the same properties but with a spin different by a unit $\frac{1}{2}$ (the mass is also different as SUSY must be broken in Nature). The contribution of the supersymmetric particles to the energy evolution of the gauge coupling constants makes that the latter can indeed meet at a single point at a scale slightly below the Planck scale; and the three interactions unify into a single one. In addition, the lightest of these new particles is the ideal candidate for dark matter in the Universe. Most important, SUSY protects the Higgs mass from acquiring large values as the quadratically divergent quantum corrections from standard particles are exactly compensated by the contributions of their supersymmetric partners. These new particles should not be heavier than 1 TeV not to spoil this compensation and, thus, they should be produced at the LHC.

---

[1] Among these, theories with extra space-time dimensions that emerge at the TeV scale and composite models inspired from the strong interactions also at the TeV scale are the most discussed ones. Some versions of these scenarios do not incorporate any Higgs particle in their spectrum and are thus ruled out by the Higgs discovery.

The Higgs discovery is particularly important for SUSY and, in particular, for its simplest low energy manifestation, the minimal supersymmetric SM (MSSM) that indeed predicts a light Higgs state with a mass below ≈ 130 GeV. In the MSSM, two Higgs doublet fields H_u and H_d are required for EWSB and to give masses to gauge bosons and (separately) to isospin up and down fermions. This leads to an extended Higgs sector compared to the SM: the physical spectrum consists of five Higgs particles, two CP-even h and H, a CP-odd A and two charged $H^\pm$ states [4, 6]. Nevertheless, because of SUSY, only two parameters are needed to describe the Higgs sector at tree-level: one Higgs mass, which is generally taken to be that of the pseudoscalar boson M_A, and the ratio of vacuum expectation values of the two Higgs fields, $\tan\beta = v_d/v_u$, expected to lie in the range $1 \lesssim \tan\beta \lesssim 60$. The masses of the CP-even h, H and the charged $H^\pm$ states, as well as the mixing angle α in the CP-even sector are uniquely defined in terms of these two inputs at tree-level, but this nice property is spoiled at higher orders [16, 17, 18, 19, 20, 21, 22].

At high M_A values, $M_A \gg M_Z$, one is in the so-called decoupling regime [23] in which the neutral CP-even state h is light and has almost exactly the properties of the SM Higgs boson, i.e., its couplings to fermions and gauge bosons are the same as the standard Higgs, while the other CP-even H and the charged $H^\pm$ bosons become heavy and mass degenerate with the A state, $M_H \approx M_{H^\pm} \approx M_A$, and decouple from the massive gauge bosons. In this regime, the MSSM Higgs sector thus looks almost exactly as the one of the SM with its unique Higgs boson.

There is, however, one major difference between the two cases: while in the SM the Higgs mass is essentially a free parameter (and should simply be smaller than about 1 TeV in order to insure unitarity in the high-energy scattering of massive gauge bosons), the lightest MSSM CP-even Higgs particle mass is bounded from above and, depending on the SUSY parameters that enter the important quantum corrections, is restricted to $M_h^{\max} \approx 90$–130 GeV. The lower value comes from experimental constraints, in particular Higgs searches at LEP [7, 8], while the upper bound assumes a SUSY breaking scale that is not too high, $M_S \lesssim \mathcal{O}$ (1 TeV), in order to avoid too much fine-tuning in the model. Hence, the requirement that the MSSM h boson coincides with the one observed at the LHC, i.e., with $M_h \approx 125$ GeV and almost SM-like couplings as the LHC data seem to indicate, would place very strong constraints on the MSSM parameters, in particular the SUSY scale M_S, through their contributions to the radiative corrections to the Higgs sector. This comes in addition to the limits that have been obtained from the search of the heavier Higgs states at the LHC, and from the negative search for superparticles.

In this review, we summarise the implications of the discovery of the H boson at the LHC and the measurement of some of its fundamental properties made so far, in particular its mass and its couplings to other particles. This will be done first in the context of the SM and then, in the MSSM. This review heavily relies on original work made by the author and published in various reviews during the last two years.

2. Implications for the standard model

2.1. Implications of the Higgs mass value

2.1.1. A triumph for the standard model. Before the advent of the LHC, electroweak precision data were collected at LEP, SLC, Tevatron and elsewhere and have provided a decisive test of the SM. These tests have been performed at the per mille level and have probed the quantum corrections of the $SU(2)_L \times U(1)_Y$ electroweak theory, establishing its gauge sector with great confidence. An important consequence of these tests has been that the only missing parameter of the model, the Higgs particle mass, was severely constrained.

As mentioned previously, taking into account all the precision electroweak data in a combined or global fit, one can indirectly constrain the Higgs boson mass M_H that contributes logarithmically to the radiative corrections to these data. The global fit of the LEP collaborations (as made just before the Higgs discovery) is displayed in the right-hand side of Figure 1, including the most up-to-date information on the other relevant parameters of the SM (in particular the top quark mass and strong coupling constant that will be discussed later). One obtains a central value [9]

$$M_H = 92^{+34}_{-26} \text{ GeV} \tag{1}$$

which corresponds to the 95% CL upper limit of $M_H \lesssim 161$ GeV [9]. This bound was pretty stable since almost two decades, when the top quark was discovered at the Tevatron and it mass was measured to be approximately $m_t \approx 173$ GeV [24].

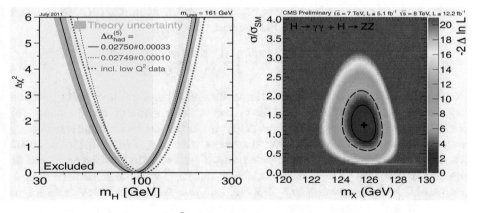

FIGURE 1. Left: the $\Delta\chi^2$ of the fit to the electroweak data as a function of M_H as performed by the LEP collaborations [9]; the solid line results when all data are included and the blue/shaded band is the estimated theoretical error from unknown higher-order corrections. Right: the Higgs mass versus the normalised production rate compared to the SM one as measured by CMS [26].

The first fundamental property of the Higgs boson that has been determined, when the particle was observed, was its mass, $M_H \approx 125$ GeV. More precisely, using the full ≈ 25 fb^{-1} data collected at $\sqrt{s}=7+8$ TeV and by ATLAS and CMS [25, 26] leads to an average value

$$M_H \simeq 125.6 \pm 0.4 \text{ GeV}. \tag{2}$$

This is remarkably close to the central value on M_H that was obtained from the fit of the precision data eq. (1); in fact, it lies with 1σ of this central value! This is a very non-trivial consistency check of the theory and it should be considered as a great achievement and a triumph for the SM of particle physics (in much the same way, and in fact even more, as the discovery of the top quark with a mass which was very close to the one predicted by the precision data, twenty years ago).

2.1.2. Stability of the electroweak vacuum. Once the Higgs boson was discovered and it mass was measured, an immediate question was whether such a Higgs boson mass value allows to extrapolate the SM up to ultimate scales, while still having an absolutely stable electroweak vacuum [13, 14] (the triviality issue, i.e., when the Higgs self-coupling becomes non-perturbative is safe, as M_H is much smaller than the limit of $M_H \approx 180$ GeV). Indeed, it is well known that top quark quantum corrections tend to drive the quartic Higgs coupling λ, which in the SM is related to the Higgs mass by the tree-level expression $\lambda = M_H^2/2v^2$ where $v \approx 246$ GeV is the Higgs field vacuum expectation value, to negative values which render the electroweak vacuum unstable.

A very recent analysis, including the state-of-the-art quantum corrections at next-to-next-to-leading order (NNLO) that are relevant in this context gives for the condition of absolute stability of the electroweak vacuum, $\lambda(M_P) \geq 0$, when the SM is extrapolated up to the Planck scale M_P [14]

$$M_H \geq 129.6 + 1.8 \times \left(\frac{m_t^{\text{pole}} - 173.2 \text{ GeV}}{0.9 \text{ GeV}} \right) - \frac{1}{2} \times \left(\frac{\alpha_s(M_Z) - 0.1184}{0.0007} \right) \pm 1.0 \text{ GeV}. \tag{3}$$

Besides the Higgs boson mass and the estimated theoretical uncertainty on the Higgs mass bound $\Delta M_H = \pm 1.0$ GeV, the vacuum stability condition eq. (3) critically depends on two basic inputs. A first one is the strong coupling constant α_s evaluated at the scale M_Z with a world average value of $\alpha_s(M_Z) = 0.1184 \pm 0.0007$ [7]. A second one is the top quark pole mass, identified with the one measured at the Tevatron by the CDF and D0 collaborations $m_t^{\text{exp}} = 173.2 \pm 0.9$ GeV [24]. A change of the input m_t value by 1 GeV will lead to a $\Delta M_H \approx \pm 2$ GeV variation of the Higgs mass bound. Allowing for a 2σ variation of m_t^{exp} alone, one obtains the upper bound $M_H \geq 125.6$ GeV that is exactly to the central value of the measured M_H eq. (2).

Thus, the "fate of the universe", i.e., whether the electroweak vacuum is stable or not up to the largest possible high-energy scale, critically relies on a precise determination (besides α_s) of M_H and m_t. This is particularly critical in the latter case as the top quark mass parameter measured at the Tevatron (and to be measured at the LHC) via kinematical reconstruction from the top quark decay products and comparison to Monte Carlo simulations, is not necessarily the pole mass which should enter the stability bound; the issue is discussed in detail in Ref. [27].

For an unambiguous and theoretically well-defined determination of the top quark mass, one should rather use the total cross section for top quark pair production at hadron colliders which can unambiguously be defined in the on-shell or $\overline{\text{MS}}$ scheme. Confronting the latest predictions of the inclusive $p\bar{p} \to t\bar{t} + X$ cross section up to next-to-next-to-leading order in QCD to the experimental measurement at the Tevatron, the running mass in the $\overline{\text{MS}}$-scheme can be determined to be $m_t^{\overline{\text{MS}}}(m_t) = 163.3 \pm 2.7$ GeV which gives a top quark pole mass of $m_t^{\text{pole}} = 171 \pm 3$ GeV.

Using these m_t and M_H inputs and adopting the value $\alpha_s = 0.1187$, the resulting contours in the $[M_H, m_t^{\text{pole}}]$ plane are confronted in Figure 2 with the areas in which the SM vacuum is absolutely stable, meta-stable and unstable up to the Planck scale.

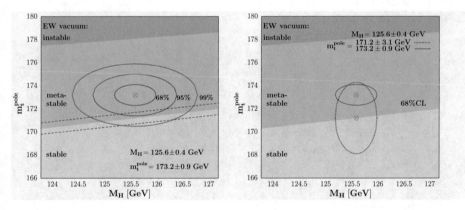

FIGURE 2. The ellipses in the $[M_H, m_t^{\text{pole}}]$ plane with the inputs $M_H = 125.6 \pm 0.4$ GeV and $\alpha_s = 0.1187$ are confronted with the areas in which the SM vacuum is absolutely stable, meta-stable and unstable up to the Planck scale. Left: the $1\sigma, 2\sigma$ and 3σ ellipses if m_t^{pole} is identified with the mass $m_t = 173.2 \pm 0.9$ GeV currently measured at the Tevatron; the black dotted lines indicate the theoretical uncertainty of $\Delta M_H = \pm 1$ GeV in the determination of the stability bound. Right: the 1σ ellipses when m_t^{pole} is identified with the one measured at the Tevatron and with the mass $m_t = 171.2 \pm 3.1$ GeV extracted for the $t\bar{t}$ production cross section.

In the left-hand side of the figure, displayed are the $68\%, 95\%$ and 99% confidence level contours if m_t^{pole} is identified with the mass measured at the Tevatron, $m_t = 173.2 \pm 0.9$ GeV. It shows that at the 2σ level, the electroweak vacuum could be absolutely stable as the ellipse almost touches the green area; this is particularly true if one includes the estimated theoretical uncertainty of $\Delta M_H = \pm 1$ GeV in the determination of the stability bound and indicated by the two black dotted lines.

In the right-hand side of Figure 2, shown are the 68%CL contours when m_t^{pole} is identified with the one measured at the Tevatron and with the mass $m_t = 171.2 \pm 3.1$ GeV extracted for the $t\bar{t}$ production cross section. In the latter case, one sees that the central value of the top mass lies almost exactly on the boundary between vacuum stability and meta-stability. The uncertainty on the top quark mass is nevertheless presently too large to clearly discriminate between these two possibilities.

2.2. Implications from the Higgs production rates

2.2.1. Light Higgs decay and production at the LHC.
In many respects, the Higgs particle was born under a very lucky star as the mass value of ≈ 125 GeV allows to produce it at the LHC in many redundant channels and to detect it in a variety of decay modes. This allows detailed studies of the Higgs properties as will be discussed in this section.

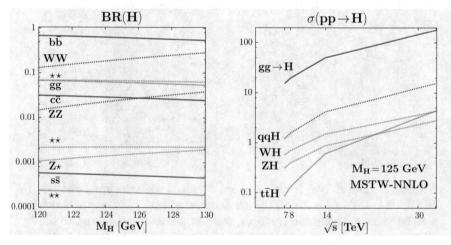

FIGURE 3. The SM-like Higgs boson branching ratios in the mass range 120–130 GeV (left) and its production cross sections at proton colliders as a function of the c.m. energy (right) [29].

We start by summarizing the production and decay at the LHC of a light SM-like Higgs particle (which should correspond to the lightest MSSM h boson in the decoupling regime). First, for $M_h \approx 125$ GeV, the Higgs mainly decays into $b\bar{b}$ pairs but the decays into WW^* and ZZ^* final states, before allowing the gauge bosons to decay leptonically $W \to \ell\nu$ and $Z \to \ell\ell$ ($\ell = e, \mu$), are also significant. The $h \to \tau^+\tau^-$ channel (as well as the gg and $c\bar{c}$ decays that are not detectable at the LHC) is also of significance, while the clean loop induced $h \to \gamma\gamma$ mode can be easily detected albeit its small rates. The very rare $h \to Z\gamma$ and even $h \to \mu^+\mu^-$ channels should be accessible at the LHC but only with a much larger data sample. This is illustrated in the left-hand side of Figure 3 where the decay branching fractions of a SM-like Higgs are displayed for the narrow mass range $M_h = 120$–130 GeV.

On the other hand, many Higgs production processes have significant cross sections as is shown in the right-hand side of Figure 3 where they are displayed at a proton collider at various past, present and foreseen center of mass energies for a 125 GeV SM-like Higgs boson; the MSTW parton densities [28] have been used.

While the by far dominant gluon fusion mechanism $gg \to h$ (ggF) has extremely large rates (≈ 20 pb at $\sqrt{s} = 7$–8 TeV), the subleading channels, i.e., the vector boson fusion (VBF) $qq \to hqq$ and the Higgs-strahlung (HV) $q\bar{q} \to hV$ with $V = W, Z$ mechanisms, have cross sections which should allow for a study of the Higgs particle already at $\sqrt{s} \gtrsim 8$ TeV with the amount of integrated luminosity, ≈ 25 fb^{-1}, that has been collected by each experiment. The Higgs-top associated process $pp \to t\bar{t}h$ (ttH) would require higher energy and luminosity.

This pattern already allows the ATLAS and CMS experiments to observe the Higgs boson in several channels and to measure some of its couplings in a reasonably accurate way. The channels that have been searched are $h \to ZZ^* \to 4\ell^{\pm}, h \to WW^* \to 2\ell 2\nu, h \to \gamma\gamma$ where the Higgs is mainly produced in ggF with subleading contributions from hjj in the VBF process, $h \to \tau\tau$ where the Higgs is produced in association with one (in ggF) and two (in VBF) jets, and finally $h \to b\bar{b}$ with the Higgs produced in the HV process. One can ignore for the moment the additional search channels $h \to \mu\mu$ and $h \to Z\gamma$ for which the sensitivity is still too low with the data collected so far.

2.2.2. The Higgs signal strengths. A convenient way to scrutinize the couplings of the produced h boson is to consider their deviation from the SM expectation. One then considers for a given search channel the signal strength modifier μ which, with some approximation, can be identified with the Higgs production cross section times decay branching fractions normalized to the SM value. For the $h \to XX$ decay channel, one would have in the narrow width approximation,

$$\mu_{XX}|_{\text{th}} = \frac{\sigma(pp \to h \to XX)}{\sigma(pp \to h \to XX)|_{\text{SM}}} = \frac{\sigma(pp \to h) \times \text{BR}(h \to XX)}{\sigma(pp \to h)|_{\text{SM}} \times \text{BR}(h \to XX)|_{\text{SM}}}, \quad (4)$$

which from the experimental side would correspond to

$$\mu_{XX}|_{\text{exp}} \simeq \frac{N_{XX}^{\text{ev}}}{\epsilon \times \sigma(pp \to h)|_{\text{SM}} \times \text{BR}(h \to XX)|_{\text{SM}} \times \mathcal{L}}, \quad (5)$$

where N_{XX}^{ev} is the measured number of events in the XX channel, ϵ the selection efficiency and $\mathcal{L}$ the luminosity.

ATLAS and CMS have provided the signal strengths for the various final states with a luminosity of, respectively, ≈ 5 fb^{-1} for the 2011 run at $\sqrt{s} = 7$ TeV

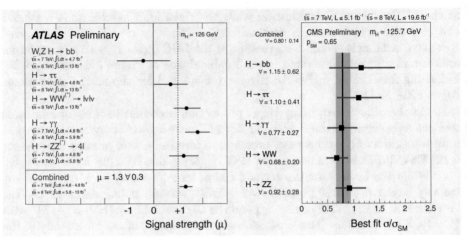

FIGURE 4. The signal strengths on the SM Higgs boson in the various search channels provided by ATLAS [30] and CMS [31] with the data collected so far at $\sqrt{s} = 7+8$ TeV.

and ≈ 20 fb^{-1} for the 2012 run at $\sqrt{s} = 8$ TeV. The constraints given by the two collaborations are shown in Figure 4.

When the various analyzed Higgs search channels are combined, this leads to a global signal strength [30, 31]

$$\text{ATLAS} : \mu_{\text{tot}} = 1.30 \pm 0.30$$
$$\text{CMS} : \mu_{\text{tot}} = 0.87 \pm 0.23 \,, \qquad (6)$$

which shows a good agreement with the SM expectation. In fact, when the ATLAS and CMS values are combined, one finds a global signal strength that is very close to unity, implying that the observed Higgs is rather SM-like.

Hence, already with the rather limited statistics at hand, the accuracy of the measurements in eq. (6) is reaching the 20% level for the ATLAS and CMS collaborations. This is at the same time impressive and worrisome. Indeed, as mentioned earlier the main Higgs production channel is the top and bottom quark loop mediated gluon fusion mechanism and, at $\sqrt{s} = 7$ or 8 TeV, the three other mechanisms contribute at a level below 15% when their rates are added and before kinematical cuts are applied. The majority of the signal events presently observed at the LHC, in particular in the main search channels $h \to \gamma\gamma, h \to ZZ^* \to 4\ell, h \to WW^* \to 2\ell 2\nu$ and, to a lesser extent $h \to \tau\tau$, thus comes from the ggF mechanism which is known to be affected by large theoretical uncertainties.

2.2.3. The theoretical uncertainties and the signal strength ratios. As a matter of fact, although the cross section $\sigma(gg \to h)$ is known up next-to-next-to-leading order (NNLO) in perturbative QCD (and at least at NLO for the electroweak interaction) [32], there is a significant residual scale dependence which points to the possibility that still higher-order contributions beyond NNLO cannot be totally excluded. In addition, as the process is of $\mathcal{O}(\alpha_s^2)$ at LO and is initiated by gluons, there are sizable uncertainties due to the gluon parton distribution function (PDF) and the value of the coupling α_s. A third source of theoretical uncertainties, the use of an effective field theory (EFT) approach to calculate the radiative corrections beyond the NLO approximation, should in principle also be considered [33, 34]. In addition, large uncertainties arise when the $gg \to h$ cross section is broken into the jet categories $h+0j, h+1j$ and $h+2j$ [35]. In total, the combined theoretical uncertainty has been estimated to be of order $\Delta^{\text{th}} \approx \pm 15\%$ by the LHC Higgs cross section working group [32] and it would increase up to $\Delta^{\text{th}} \approx \pm 20\%$ if the EFT uncertainty is also included[2] [34].

Hence, the theoretical uncertainty is already at the level of the accuracy of the cross section measured by the ATLAS and CMS collaborations, eq. (6). Another drawback of the analyses is that they involve strong theoretical assumptions on the total Higgs width since some contributing decay channels not accessible at the LHC are assumed to be SM-like and possible invisible Higgs decays in scenarios beyond the SM are supposed not to occur.

In Ref. [29], following earlier work [36], it has been suggested to consider the decay ratios D_{XX} defined as

$$D_{XX}^{\text{p}} = \frac{\sigma^{\text{P}}(pp \to h \to XX)}{\sigma^{\text{P}}(pp \to h \to VV)} = \frac{\sigma^{\text{P}}(pp \to h) \times \text{BR}(h \to XX)}{\sigma^{\text{P}}(pp \to h) \times \text{BR}(h \to VV)} = \frac{\Gamma(h \to XX)}{\Gamma(h \to VV)} \propto \frac{c_X^2}{c_V^2} \tag{7}$$

for a specific production process p = ggF, VBF, VH or all (for inclusive production) and for a given decay channel $h \to XX$ when the reference channel $h \to VV$ is used. In these ratios, the cross sections $\sigma^P(pp \to h)$ and hence, their significant theoretical uncertainties will cancel out, leaving out only the ratio of decay branching fractions and hence of partial decay widths. These can be obtained with the program HDECAY [37] which includes all higher-order effects and are affected by much smaller uncertainties. Thus, the total decay width which includes contributions from channels not under control such as possible invisible Higgs decays, do not appear in the ratios D_{XX}^{p}. Some common experimental systematical uncertainties such as the one from the luminosity measurement and the small uncertainties in the Higgs decay branching ratios also cancel out. We are thus, in principle, left with only the statistical uncertainty and some (non common) systematical errors. The ratios D_{XX} involve, up to kinematical factors and known radiative corrections, only the ratios $|c_X|^2/ |c_V|^2$ of the Higgs reduced couplings to the particles X and V compared to the SM expectation, $c_X \equiv g_{hXX}/g_{hXX}^{\text{SM}}$.

---

[2] Note that also in the VBF process, despite the fact that the inclusive cross section has only a few percent combined scale and PDF uncertainty [32], the contamination by the $gg \to h+2j$ channel makes the total uncertainty in the $h+jj$ final "VBF" sample rather large. Indeed $\mathcal{O}(30\%)$ $gg \to h+2j$ events will remain even after the specific cuts that select the VBF configuration have been applied, and the rate is affected by a much larger uncertainty than the inclusive $gg \to h$ process, up to 50% when one adds the scale and PDF uncertainties [35].

For the time being, three independent ratios can be considered: $D_{\gamma\gamma}, D_{\tau\tau}$ and D_{bb}. $D_{\gamma\gamma}$ is the ratio of the inclusive ATLAS and CMS di-photon and ZZ channels that are largely dominated by the ggF mode; $D_{\tau\tau}$ is the signal strength ratio in the $\tau\tau$ and WW searches where one selects Higgs production in ggF with an associated jet or in the VBF production mechanism; D_{bb} is the ratio of the $h \to b\bar{b}$ and $h \to WW$ decays in hV production for which the sensitivities are currently too low.

In order to test the compatibility of the couplings of the $M_h = 125$ GeV Higgs state with the SM expectation, one can perform a fit based on the χ_R^2 function

$$\chi_R^2 = \frac{[D_{\gamma\gamma}^{ggF} - \frac{\mu_{\gamma\gamma}}{\mu_{ZZ}}|ggF|_{\exp}]^2}{[\delta(\frac{\mu_{\gamma\gamma}}{\mu_{ZZ}})_{ggF}]^2} + \frac{[D_{bb}^{VH} - \frac{\mu_{bb}}{\mu_{WW}}|Vh|_{\exp}]^2}{[\delta(\frac{\mu_{bb}}{\mu_{WW}})_{Vh}]^2} + \frac{[D_{\tau\tau}^{ggF+VBF} - \frac{\mu_{\tau\tau}}{\mu_{WW}}|ggF+VBF|_{\exp}]^2}{[\delta(\frac{\mu_{\tau\tau}}{\mu_{WW}})_{ggF+VBF}]^2}.$$

$$(8)$$

The errors $\delta(\mu_{XX}/\mu_{VV})$ are computed assuming no correlations between the different final state searches. The uncertainties on the ratios are derived from the individual errors that are dominated by the experimental ones as one expects that the theoretical uncertainties largely cancel out in the ratios $D_{\gamma\gamma}$, D_{bb} and $D_{\tau\tau}$.

For the signal strengths above, the theoretical uncertainties have to be treated as a bias (and not as if they were associated with a statistical distribution) and the fit has to be performed for the two extremal values of the signal strengths: $\mu_i|_{\exp} \pm \delta\mu_i/\mu_i|_{th}$ with the theoretical uncertainty $\delta\mu_i/\mu_i|_{th}$ conservatively assumed to be $\pm 20\%$ for both the ggF and VBF mechanisms (because of the contamination due to $gg \to h + 2j$ in the latter case) and $\approx 5\%$ for hV associated production.

2.2.4. Fit of the Higgs couplings and their ratios. A large number of analyses of the Higgs couplings from the LHC data have been performed in the SM and its extensions and a partial list is given in Refs. [38, 39, 40, 41]. In most cases, it is assumed that the couplings of the Higgs boson to the massive W, Z gauge bosons are equal to $g_{HZZ} = g_{HWW} = c_V$ and the couplings to all fermions are also the same $g_{Hff} = c_f$. However, while the first assumption is justified by custodial symmetry which holds in almost all situations, the second one is rather crude as, at least the isospin up and isospin down type couplings should be different (this holds in the case of the MSSM for instance as will be discussed later).

Hence, to characterize the Higgs particle at the LHC, at least three independent h couplings should be considered, namely c_t, c_b and $c_V = c_V^0$ as for instance advocated in Ref. [41]. One can thus define the following effective Lagrangian,

$$\mathcal{L}_h = c_V g_{hWW} h W_\mu^+ W^{-\mu} + c_V g_{hZZ} h Z_\mu^0 Z^{0\mu}$$

$$- c_t y_t h \bar{t}_L t_R - c_t y_c h \bar{c}_L c_R - c_b y_b h \bar{b}_L b_R - c_b y_\tau h \bar{\tau}_L \tau_R + \text{h.c.},$$

$$(9)$$

where $y_{t,c,b,\tau} = m_{t,c,b,\tau}/v$ are the Yukawa couplings of the heavy SM fermions, $g_{hWW} = 2M_W^2/v$ and $g_{hZZ} = M_Z^2/v$ the hWW and HZZ couplings and v the SM Higgs vev.

While the couplings to gauge bosons and bottom quarks are derived directly by considering the decays of the Higgs bosons to WW/ZZ and $b\bar{b}$ final states, the

$ht\bar{t}$ coupling is derived indirectly from the $gg \to h$ production cross section and the $h \to \gamma\gamma$ decay branching ratio, two processes that are generated by triangular loops involving the top quarks in the SM. We will assume, in a first approximation, that the couplings to charm quarks and tau leptons are such that $c_c = c_t$ and $c_\tau = c_b$; another caveat is due to possible invisible Higgs decays which are assumed to be absent; these two issues will be discussed later in the context of the MSSM.

In Ref. [41], a three-dimensional fit of the h couplings was performed in the space $[c_t, c_b, c_V]$, when the theory uncertainty is taken as a bias and not as a nuisance. The results of this fit are presented in Figure 5 (left) for $c_t, c_b, c_V \geq 0$. The best-fit value for the couplings, with the $\sqrt{s} = 7{+}8$ TeV ATLAS and CMS data turns out to be $c_t = 0.89$, $c_b = 1.01$ and $c_V = 1.02$, ie very close to the SM values.

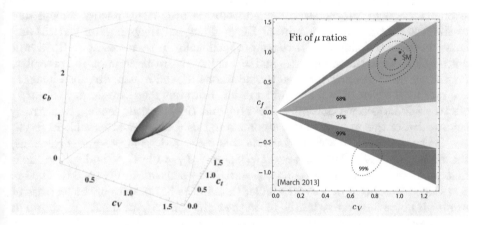

FIGURE 5. Left: The best-fit region at 68%CL for the Higgs signal strengths in the $[c_t, c_b, c_V]$ space [41]. The three overlapping regions are for the central and extreme choices of the theoretical prediction for the Higgs rates including uncertainties. Right: Best-fit regions at 68.27%CL (green), 95.45%CL (yellow) and 99.73%CL (grey) in the plane c_f versus c_V; the associated best-fit point (cross) and SM (red) point are also shown [40].

In the right-hand side of Figure 5, we assume a universal Higgs coupling to fermions $c_t = c_b = c_f$ and show the results when fitting the signal strengths as well as the Higgs decay ratios through the function χ_R^2 eq.(8) One sees that the best-fit domains from the ratios obtained, e.g., at 1σ do not exclude parts of the 1σ regions obtained from χ^2 since the main theoretical uncertainty cancels out in the D_{XX} ratios and is negligible for the signal strengths when added in quadrature to the experimental error. The domains from χ^2 are more restricted as (i) this function exploits the full experimental information on the Higgs rates and not only on the ratios and (ii) the experimental error on a ratio of rates is higher than on the rates alone. The situation might improve in the future when the experimental uncertainty will decrease.

2.2.5. Implications for a fourth generation fermions. One of the immediate implications of the LHC Higgs results is that a fourth generation of fermions is now completely ruled out [42, 43]. This SM extension that we denote SM4 is straightforward as one simply needs to add to the known fermionic pattern with three generations, two quarks t' and b' with weak-isospin $\frac{1}{2}$ and $-\frac{1}{2}$, a charged lepton ℓ' and a neutrino ν'. SM4 was still viable provided that the neutrino should be rather heavy, $m_{\nu'} \gtrsim \frac{1}{2}M_Z$, as required by the invisible width of the Z boson measured at LEP as well as $m_{\ell'} \gtrsim 100$ GeV for the charged lepton [7, 8].

In addition to the direct LHC searches that excluded too light fourth generation quarks with masses close to the unitarity bound, $m_{b'}, m_{t'} \lesssim 600$ GeV [44], strong constraints can be also obtained from the LHC Higgs results. This is due to the fact that in the loop induced Higgs-gluon and Higgs-photon vertices any heavy particle coupling to the Higgs proportionally to its mass, as in SM4, will not decouple from the amplitudes and would have a drastic impact. In particular, for the $gg \to H$ leading process, the additional t' and b' contributions increase the rate by a factor of ≈ 9 at leading order. However, there are large $\mathcal{O}(G_F m_{f'}^2)$ electroweak corrections that affect the Hgg and $H\gamma\gamma$ vertices, leading to a strong suppression of the $gg \to H \to \gamma\gamma$ rate, making this channel unobservable in SM4.

Using a version of HDECAY for SM4 that include these important corrections, the rate $\sigma(gg \to H) \times \mathrm{BR}(H \to \gamma\gamma)|_{\mathrm{SM4/SM}}$ for $M_H = 125$ GeV is displayed as a function of $m_{\nu'} = m_{\ell'}$ for the value $m_{b'} = m_{t'} + 50 = 600$ GeV in Figure 6 (left). One sees that it is a factor of 5 to 10 smaller than in the SM despite of the increase of $\sigma(gg \to H)$ by a factor of ≈ 9 in SM4. In the right-hand side of the figure, shown in the ratio $\sigma(q\bar{q} \to VH) \times \mathrm{BR}(H \to b\bar{b})|_{\mathrm{SM4/SM}}$ in the same SM4 configuration and one sees that there is a reduction of the $Vb\bar{b}$ signal rate by a factor 3 to 5 depending on $m_{\nu'}$. Hence in SM4, the Higgs signal would have not been observable in SM4 and the obtained results unambiguously rule out this possibility.

2.3. Total width and invisible Higgs decays

Invisible decays, if present will affect the properties of the observed h particle. These decays could be constrained if the total decay width could be determined. A direct measurement of the total decay width of the Higgs particle Γ_H^{tot} would have been possible if the Higgs mass were larger than $M_H \gtrsim 200$ GeV, by exploiting the process $H \to ZZ \to 4\ell^{\pm}$: beyond this mass value, $\Gamma_H^{\mathrm{tot}} \gtrsim 1$ GeV, would have been large enough to be resolved experimentally, in contrast to the width of a 125 GeV particle which in the SM is too small, $\Gamma_H^{\mathrm{tot}} = 4$ MeV.

Rather recently, it has been noticed [45] that in the channel $pp \to VV \to 4f$, a large fraction ($\approx 10\%$) of the Higgs-mediated cross section lies in the high-mass tail where the invariant mass of the VV system is larger than $2M_V$. The tail depends on the Higgs couplings that enter the production and the decay processes but not on the Higgs total width. This feature has been used by many authors in

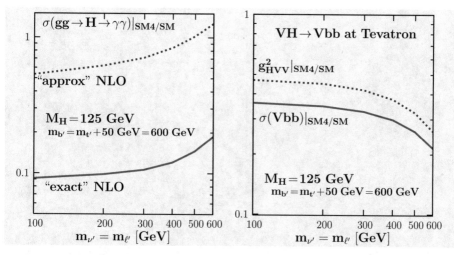

FIGURE 6. Left: $\sigma(gg \to H) \times \mathrm{BR}(H \to \gamma\gamma)|_{\mathrm{SM4/SM}}$ for a 125 GeV Higgs boson as a function of $m_{\nu'} = m_{\ell'}$ when the leading $\mathcal{O}(G_F m_{f'}^2)$ electroweak corrections are included in a naive way ("approx" NLO) or in a way that mimics the exact NLO results ("exact" NLO). Right: the HVV coupling squared and $\sigma(q\bar{q} \to VH) \times \mathrm{BR}(H \to b\bar{b})$ in SM4 normalized to the SM values. The program HDECAY for SM4 has been used; from [42].

order to constrain the total width Γ_H^{tot} [46] and recent measurements of the ATLAS and CMS collaborations in the channel $pp \to H \to ZZ^* \to 4\ell^\pm$ led to a bound $\Gamma_H^{\mathrm{tot}}/\Gamma_H^{\mathrm{SM}} \approx 5\text{--}10$ [47]. However, these bounds strongly rely on the assumption that the off-shell Higgs couplings are exactly the same as the on-shell couplings, which has been shown not to be the case in many situations [48].

Nevertheless, the invisible Higgs decay width can be constrained indirectly by a fit of the Higgs couplings and in particular with the signal strength μ_{ZZ} which is the most accurate one and has the least theoretical ambiguities. Γ_H^{inv} enters in the signal strength through the total width Γ_H^{tot}, $\mu_{ZZ} \propto \Gamma(H \to ZZ)/\Gamma_H^{\mathrm{tot}}$ with $\Gamma_H^{\mathrm{tot}} = \Gamma_H^{\mathrm{inv}} + \Gamma_H^{\mathrm{SM}}$ and Γ_H^{SM} calculated with free coefficients c_f and c_V. The resulting 1σ or 2σ ranges are shown in Figure 7 (left) where c_f is freely varied while $c_V = 1$ [40]. This gives $\Gamma_H^{\mathrm{inv}}/\Gamma_H^{\mathrm{SM}} \lesssim 50\%$ at the 95% CL if the assumption $c_f = c_V = 1$ is made.

A more model independent approach would be to perform direct searches for missing transverse energy. These have been conducted by ATLAS [49] and CMS [50] in the $pp \to hV$ process with $V \to jj, \ell\ell$ and in the VBF channel, $qq \to qq E\!\!\!/_T$. As an example, we show in Figure 7 (center) the CMS results for the Higgs cross section times $\mathrm{BR_{inv}}$ versus M_h when the two channels are combined. For $M_h \approx 125$ GeV a bound $\mathrm{BR_{inv}} \lesssim 50\%$ is obtained at the 95%CL.

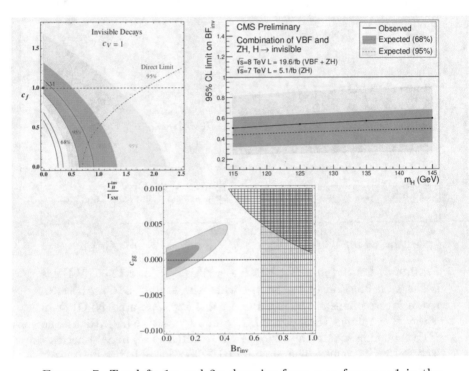

FIGURE 7. Top left: 1σ and 2σ domains from μ_{ZZ} for $c_V = 1$ in the plane $[c_f, \Gamma_H^{inv}/\Gamma_H^{tot}]$ [40]; the dependence on the theory uncertainties are shown by the black curves that indicate the other possible extreme domains and the direct upper limit on Γ_H^{inv} from direct searches at LHC for $c_V = c_f = 1$ [49] is also shown. Top right: the Higgs cross section times invisible Higgs decay branching ratio normalised to the total SM cross section in the combined hV and VBF channels from CMS with the ≈ 20 fb^{-1} data at 8 TeV [50]. Bottom: 68% CL (light green) and 95% CL (dark green) best fit regions to the combined LHC Higgs data. The black region is excluded by the monojet constraints while the red region is excluded by the ATLAS $Z + E_T$ search [49]; from Ref. [51].

A more promising search for invisible decays is the monojet channel. In the ggF mode, an additional jet can be emitted at NLO leading to $gg \to hj$ final states and, because the QCD corrections are large, $\sigma(H + 1j)$ is not much smaller than $\sigma(h + 0j)$. The NNLO corrections besides significantly increasing the $h + 0j$ and $h + 1j$ rates, lead to $h + 2j$ events that also occur in VBF and VH. Hence, if the Higgs is coupled to invisible particles, it may recoil against hard QCD radiation, leading to monojets.

In Refs. [51, 52], it has been shown that the monojet signature carries a good potential to constrain the invisible decay width of a ≈ 125 GeV Higgs boson.

In a model independent fashion, constraints can be placed on $R_{\text{inv}}^{\text{ggF}} = \sigma(gg \rightarrow h) \times \text{BR}(h \rightarrow \text{inv.}) / \sigma(gg \rightarrow h)_{\text{SM}}$. Recent monojet searches made by CMS and ATLAS [53] are sensitive to R_{inv} close to unity. This is shown in Figure7 (right) where the best fit region to the LHC Higgs data is displayed in the Br_{inv}-c_{gg} plane, where c_{gg} is the deviation of $\sigma(gg \rightarrow h)$ from the SM expectation [51]. For the SM value $c_{gg} = 0$, $\text{Br}_{\text{inv}} \gtrsim 20\%$ is disfavored at 95% CL while for $c_{gg} > 0$, a rate up to 50% is allowed.

The Higgs invisible rate and the dark matter detection rate in direct astro-physical searches are correlated in Higgs portal models. Considering the generic cases of scalar, fermionic and vectorial dark matter particles χ that couple only to the Higgs, one can translate in each case the LHC constraint $\text{BR}(h \rightarrow \text{inv.})$ into a constraint on the Higgs couplings to the χ particles. It turns out that these constraints are competitive with those derived from the bounds on the dark matter scattering cross section on nucleons from the best experiment so far, XENON [54].

This is shown in Figure 8 where the maximum allowed values of the scattering cross sections are given in the three cases assuming $\text{BR}_{\chi}^{\text{inv}} \lesssim 20\%$ [55]. The obtained spin-independent rates $\sigma_{\chi p}^{\text{SI}}$ are stronger than the direct limit from the XENON100 experiment in the entire $M_{\chi} \ll \frac{1}{2} M_h$ range. In other words, the LHC is currently the most sensitive dark matter detection apparatus, at least in the context of simple Higgs-portal models.

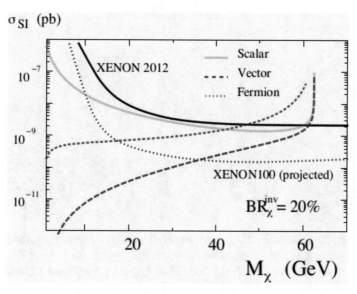

FIGURE 8. Bounds on the spin-independent direct detection cross section $\sigma_{\chi p}^{\text{SI}}$ in Higgs portal models derived for an invisible branching fraction of 20 % (colored lines) for a 125 GeV Higgs. These are compared to current and future direct bounds from XENON experiment (black). From Ref. [55].

2.4. Determination of the Higgs parity

Apart from the measurement of the couplings, one also needs in principle to establish that the observed Higgs state is indeed a CP even scalar particle and hence with $J^{PC} = 0^{++}$ quantum numbers[3]. It is known that the Higgs to vector boson (hVV) coupling is a possible tool to probe these quantum numbers at the LHC [58]. This can be done by studying various kinematical distributions in the Higgs decay and production processes. One example is the threshold behavior of the M_{Z^*} spectrum in the $h \to ZZ^* \to 4\ell$ decay channel and another is the azimuthal distribution between the decay planes of the two lepton pairs arising from the Z, Z^* bosons from the Higgs decay. These are sensitive to both the spin and parity of the Higgs.

With the 25 fb^{-1} data collected so far, the ATLAS and CMS collaborations performed a matrix-element likelihood analysis which exploits the kinematics and Lorenz structure of the $h \to ZZ^* \to 4\ell$ channel to see whether the angular distributions are more compatible with the 0^+ or 0^- hypothesis (as well as the spin-2 possibility) [59]. Assuming that it has the same couplings as the SM Higgs boson and that it is produced mainly from the dominant ggF process, the observed particle is found to be compatible with a 0^+ state and the 0^- possibility is excluded at the 97.8% confidence level or higher; see Figure 9.

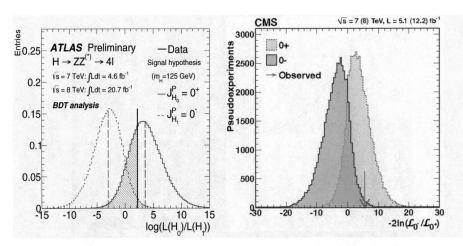

FIGURE 9. Discrimination between the 0^+ and 0^- parity hypotheses for the observed Higgs boson using the kinematics of the $h \to ZZ^* \to 4\ell$ channel by the ATLAS (left) and CMS (right) collaborations with the data collected at 7+8 TeV [59].

---

[3]To be more general, the spin of the particle needs also to be determined. The observation of the $h \to \gamma\gamma$ decay channel rules out the spin-1 case by virtue of the Landau–Yang theorem [56] and leaves only the spin 0 and ≥ 2 possibilities. The graviton-like spin-2 option is extremely unlikely and, already from the particle signal strengths in the various channels, it is ruled out in large classes of beyond the SM models; see, e.g., Ref. [57].

Other useful diagnostics of the CP nature of the Higgs boson that also rely on the different tensorial structure of the hVV coupling can be made in the VBF process. It was known since a long time that in this channel, the distribution in the azimuthal angle between the two jets produced in association with the Higgs discriminates a CP-even from a CP-odd state [60]. This has been extended recently to other observables, like the rapidity separation between the two jets [61, 62].

Recently, the VBF channel $pp \rightarrow Hjj$ has been reanalyzed in the presence of an anomalous hVV vertex that parametrises different spin and CP assignments of the produced Higgs boson [62]. The anomalous hVV coupling is introduced by allowing for an effective Lagrangian with higher-dimensional operators, that include four momentum terms which are absent in the SM. It was shown that the kinematics of the forward tagging jets in this process is highly sensitive to the structure of the anomalous coupling and that it can effectively discriminate between different assignments for the spin (spin-0 versus spin-2) and the parity (CP-even versus CP-odd) of the produced particle. In particular, it was found that the correlation between the separation in rapidity and the transverse momenta of the scattered quarks, in addition to the already discussed distribution of the azimuthal jet separation, can be significantly altered compared to the SM expectation.

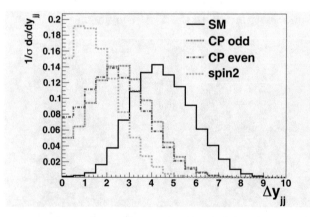

FIGURE 10. Normalized distribution of the difference in rapidity between the scattered jets in VBF for each of the SM and BSM operators (spin-2, CP-even and CP-odd) individually [62].

Figure 10 shows the difference in rapidity between tagging jets (Δy_{jj}) for each of the higher-dimensional operators in the hVV couplings.

These kinematical variables define new corners of the phase-space that have not been explored by the experiments at the LHC to probe anomalous hVV couplings and to check the Higgs parity. In addition, some of these observables significantly depend on the c.m. energy and strong constraints on anomalous couplings can be obtained by performing measurements at the LHC with energies of $\sqrt{s}=8$

TeV and 14 TeV. Finally, the associated hV production channel can be used as the invariant mass of the Vh system as well as the p_T and rapidities of the h and V bosons are also sensitive to anomalous hVV couplings.

Nevertheless, there is a caveat in the analyses relying on the hVV couplings. Since a CP-odd state has no tree-level VV couplings, all the previous processes project out only the CP-even component of the hVV coupling [63] even if the state is a CP-even and odd mixture. Thus, in the CP studies above, one is simply verifying a posteriori that indeed the CP-even component is projected out.

A better way to measure the parity of the Higgs boson is to study the signal strength in the $h \rightarrow VV$ channels [40, 64]. Indeed, the hVV coupling takes the general form $g_{hVV}^{\mu\nu} = -ic_V(M_V^2/v)\, g^{\mu\nu}$ where c_V measures the departure from the SM: $c_V = 1$ for a pure 0^+ state with SM-like couplings and $c_V \approx 0$ for a pure 0^- state. The measurement of c_V should allow to determine the CP composition of the Higgs if it is indeed a mixture of 0^+ and 0^- states.

However, having $c_V \neq 1$ does not automatically imply a CP-odd component: the Higgs sector can be enlarged to contain other states h_i with squared h_iVV couplings $\Sigma_i c_{V_i}^2\, g_{h_iVV}^2$ that reduce to the SM coupling g_{hVV}^2. This is what occurs in the MSSM with complex soft parameters [58]: one has three neutral states h_1, h_2 and h_3 with indefinite parity and their CP-even components share the SM hVV coupling, $c_{V_1}^2 + c_{V_2}^2 + c_{V_3}^2 = 1$. But in all cases, the quantity $1 - c_V^2$ gives an *upper bound* on the CP-odd contribution to the hVV coupling.

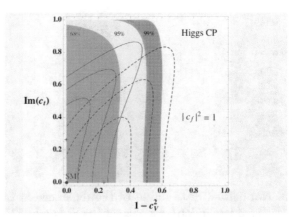

FIGURE 11. Best-fit regions at 68%, 95% and 99%CL in the plane $[1 - c_V^2, \mathrm{Im}(c_t)$ for $|c_t|^2 = |c_f|^2 = 1$. Superimposed are the best-fit regions when including a theory uncertainty of $\pm 20\%$ [40].

Using μ_{VV} and the ratios $\mu_{\gamma\gamma}/\mu_{VV}$ and $\mu_{\tau\tau}/\mu_{VV}$ as in eq. (8), it was demonstrated that the particle has indeed a large CP component, $\gtrsim 50\%$ at the 95%CL, if the Higgs couplings to fermions are SM like. This is shown in Figure 11 where one sees that the pure CP-odd possibility is excluded at the 3σ level, irrespective of the (mixed CP) Higgs couplings to fermions provided that $|c_f|^2 = 1$.

3. Implications for supersymmetry

We turn now to the implications of the LHC Higgs results for the MSSM Higgs sector from the mass value, the production rates and from the search for the heavy Higgs bosons. However, we first discuss a few basics issues concerning the Higgs masses and couplings when the important radiative corrections are taken into account.

3.1. The Higgs masses and couplings in the MSSM

In the MSSM, the tree-level masses of the CP-even h and H bosons depend only on M_A and $\tan\beta$. However, many parameters of the MSSM such as the masses of the third generation stop and sbottom squarks $m_{\tilde{t}_i}, m_{\tilde{b}_i}$ and their trilinear couplings A_t, A_b enter M_h and M_H through quantum corrections. In the basis (H_d, H_u), the CP-even Higgs mass matrix can be written in full generality as

$$\mathcal{M}^2 = M_Z^2 \begin{pmatrix} c_\beta^2 & -s_\beta c_\beta \\ -s_\beta c_\beta & s_\beta^2 \end{pmatrix} + M_A^2 \begin{pmatrix} s_\beta^2 & -s_\beta c_\beta \\ -s_\beta c_\beta & c_\beta^2 \end{pmatrix} + \begin{pmatrix} \Delta\mathcal{M}_{11}^2 & \Delta\mathcal{M}_{12}^2 \\ \Delta\mathcal{M}_{12}^2 & \Delta\mathcal{M}_{22}^2 \end{pmatrix},$$
(10)

where we use the short-hand notation $s_\beta \equiv \sin\beta$ etc... and introduce the radiative corrections by a general 2×2 matrix $\Delta\mathcal{M}_{ij}^2$. One can then easily derive the neutral CP even Higgs boson masses and the mixing angle α that diagonalises the h and H states, $H = \cos\alpha H_d^0 + \sin\alpha H_u^0$ and $h = -\sin\alpha H_d^0 + \cos\alpha H_u^0$:

$$M_{h/H}^2 = \frac{1}{2}\left(M_A^2 + M_Z^2 + \Delta\mathcal{M}_+^2 \mp N\right),$$
(11)

$$\tan\alpha = \frac{2\Delta\mathcal{M}_{12}^2 - (M_A^2 + M_Z^2)s_\beta}{\Delta\mathcal{M}_-^2 + (M_Z^2 - M_A^2)c_{2\beta} + N},$$
(12)

$$\Delta\mathcal{M}_\pm^2 = \Delta\mathcal{M}_{11}^2 \pm \Delta\mathcal{M}_{22}^2, \quad N = \sqrt{M_A^4 + M_Z^4 - 2M_A^2 M_Z^2 c_{4\beta} + C},$$

$$C = 4\Delta\mathcal{M}_{12}^4 + (\Delta\mathcal{M}_-^2)^2 - 2(M_A^2 - M_Z^2)\Delta\mathcal{M}_-^2 c_{2\beta} - 4(M_A^2 + M_Z^2)\Delta\mathcal{M}_{12}^2 s_{2\beta}.$$

The by far leading one-loop radiative corrections to the mass matrix of eq. (10) are controlled by the top Yukawa coupling, $\lambda_t = m_t/v\sin\beta$ with $v = 246$ GeV, which appears with the fourth power. One obtains a very simple analytical expression for the correction matrix $\Delta\mathcal{M}_{ij}^2$ if only this contribution is taken into account [16],

$$\Delta\mathcal{M}_{11}^2 \sim \Delta\mathcal{M}_{12}^2 \sim 0,$$
(13)

$$\Delta\mathcal{M}_{22}^2 \sim \epsilon = \frac{3\bar{m}_t^4}{2\pi^2 v^2 \sin^2\beta}\left[\log\frac{M_S^2}{\bar{m}_t^2} + \frac{X_t^2}{M_S^2}\left(1 - \frac{X_t^2}{12M_S^2}\right)\right],$$

where M_S is the geometric average of the two stop masses $M_S = \sqrt{m_{\tilde{t}_1} m_{\tilde{t}_2}}$ defined to be the SUSY-breaking scale and X_t is the stop mixing parameter given by $X_t = A_t - \mu/\tan\beta$ with μ the higgsino mass parameter; $\bar{m}_t$ is the running $\overline{\text{MS}}$ top quark mass to account for the leading two-loop QCD corrections in a renormalisation-group improved approach (some refinements can be included as well).

Other soft SUSY-breaking parameters, in particular μ and A_b (and in general the corrections controlled by the bottom Yukawa coupling $\lambda_b = m_b/v\cos\beta$ which at large value of $\mu \tan\beta$ become relevant) as well as the gaugino mass parameters $M_{1,2,3}$, provide a small but non-negligible correction to $\Delta\mathcal{M}^2_{ij}$ and can thus also have an impact on the loop corrections [17, 20, 21, 22].

The maximal value M_h^{max} is given in the leading one-loop approximation by

$$M_h^2 \overset{M_A \gg M_Z}{\to} M_Z^2 \cos^2 2\beta + \Delta\mathcal{M}^2_{22} \tag{14}$$

and is obtained for the choice of parameters [20, 21, 22]: i) a decoupling regime with heavy A states, $M_A \sim \mathcal{O}(\mathrm{TeV})$; ii) large values of the parameter $\tan\beta$, $\tan\beta \gtrsim 10$; iii) heavy stops, i.e., large M_S values and we choose in general $M_S \leq 3$ TeV to avoid a too large fine-tuning [65, 66]; iv) a stop trilinear coupling $X_t = \sqrt{6}M_S$, the so-called maximal mixing scenario that maximizes the stop loops [67]. If the parameters are optimized as above, the maximal M_h value can reach the level of 130 GeV.

An important aspect is that in the decoupling regime $M_A \gg M_Z$, the heavier CP-even and the charged Higgs states become almost degenerate in mass with the CP-odd state, $M_H \approx M_{H^\pm} \approx M_A$, while the mixing angle α becomes close to $\alpha \approx \frac{\pi}{2} - \beta$ making the couplings of the light h state to fermions and massive gauge bosons SM-like, and decoupling the $H, H^\pm$ from the weak bosons as is the case for the state A by virtue of CP invariance.

It was pointed out in Refs. [68, 69, 41] that when the measured value $M_h = 125$ GeV is taken into account, the MSSM Higgs sector with solely the dominant radiative corrections included, can be again described with only two free parameters such as $\tan\beta$ and M_A as it was the case at tree-level. In other words, the dominant radiative corrections that involve the SUSY parameters are fixed by the value of M_h. This observation leads to a rather simple parametrisation of the MSSM Higgs sector.

More specifically, let us assume that in the 2×2 matrix for the radiative corrections to the CP-even Higgs mass matrix eq. (10), only the leading $\Delta\mathcal{M}^2_{22}$ entry of eq. (13) that involves the by far dominant stop-top sector contribution is taken into account; this is the so-called ϵ approximation and its refinements [17, 22]. In this $\Delta\mathcal{M}^2_{22} \gg \Delta\mathcal{M}^2_{11}, \Delta\mathcal{M}^2_{12}$ limit, one can simply trade $\Delta\mathcal{M}^2_{22}$ for the by now known h mass value $M_h = 125$ GeV and obtain

$$M_H^2 = \frac{(M_A^2 + M_Z^2 - M_h^2)(M_Z^2 c_\beta^2 + M_A^2 s_\beta^2) - M_A^2 M_Z^2 c_{2\beta}^2}{M_Z^2 c_\beta^2 + M_A^2 s_\beta^2 - M_h^2},$$

$$\alpha = -\arctan\left(\frac{(M_Z^2 + M_A^2)c_\beta s_\beta}{M_Z^2 c_\beta^2 + M_A^2 s_\beta^2 - M_h^2}\right). \tag{15}$$

This was called the habemus MSSM or hMSSM in Ref. [41].

However, this interesting and simplifying feature has to been demonstrated for all MSSM parameters and, in particular, one needs to prove that the impact of the subleading corrections $\Delta\mathcal{M}^2_{11}$ and $\Delta\mathcal{M}^2_{12}$ is small. To do so, a scan of the pMSSM parameter space using the program SuSpect, in which the full two-loop radiative corrections to the Higgs sector are implemented, has been performed

[41]. For a chosen $(\tan\beta, M_A)$ input set, the soft-SUSY parameters that play an important role in the Higgs sector are varied in the following ranges: $|\mu| \leq 3$ TeV, $|A_t, A_b| \leq 3M_S$, 1 TeV $\leq M_3 \leq 3$ TeV and 0.5 TeV $\leq M_S \leq 3$ TeV (≈ 3 TeV is the scale up to which programs such as SuSpect are expected to be reliable). The usual GUT relation between the weak scale gaugino masses $6M_1 = 3M_2 = M_3$ has been assumed and $A_u, A_d, A_\tau = 0$ has been set (these last parameters have little impact on the radiative corrections). The MSSM Higgs sector parameters have been computed all across the parameter space, selecting the points which satisfy the constraint $123 \leq M_h \leq 129$ GeV when uncertainties are included. For each of theses points, the Higgs parameters have been compared to those obtained in the simplified MSSM approximation, $\Delta\mathcal{M}^2_{11,12} = 0$, with the lightest Higgs boson mass as input. While the requirement that M_h should lie in the range 123–129 GeV has been made, M_h was allowed to be different from the one obtained in the "exact" case $\Delta\mathcal{M}^2_{11,12} \neq 0$.

Displayed in Figure 12 are the differences between the values of the mass M_H and the mixing angle α that are obtained when the two possibilities $\Delta\mathcal{M}^2_{11,12} = 0$ and $\Delta\mathcal{M}^2_{11}, \Delta\mathcal{M}^2_{12} \neq 0$ are considered. This is shown in the plane $[M_S, X_t]$ with $X_t = A_t - \mu\cot\beta$ when all other parameters are scanned as above. The A boson mass was fixed to $M_A = 300$ GeV (a similar result was obtained for $M_A \approx 1$ TeV) and two representative values $\tan\beta = 5$ and 30 are used. The conservative approach of plotting only points which maximize these differences has been adopted.

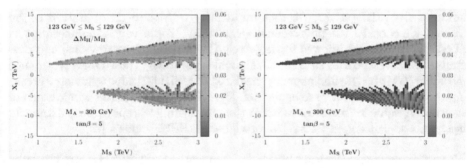

FIGURE 12. The variation of the mass M_H and the mixing angle α are shown as separate vertical colored scales, in the plane $[M_S, X_t]$ when the full two loop corrections are included with and without the subleading matrix elements $\Delta\mathcal{M}^2_{11}$ and $\Delta\mathcal{M}^2_{12}$. $M_A = 300$ GeV, $\tan\beta = 5$ are taken and the other parameters are varied as described in the text [41].

In all cases, the difference between the two M_H values is very small (in fact, much smaller than the H boson total decay width Γ_H), less than a few percent, while for α the difference does not exceed ≈ 0.025 for low values of $\tan\beta$ but at high $\tan\beta$ values, one can reach the level of ≈ 0.05 in some rare situations (large values of μ, which enhance the $\mu\tan\beta$ contributions). Nevertheless, at high enough

$\tan\beta$, we are far in the decoupling regime already for $M_A \gtrsim 200$ GeV and such a difference does not significantly affect the couplings of the h and H bosons which, phenomenologically, are the main ingredients.

Hence, even when including the full set of radiative corrections, it remains a good approximation to use eqs. (15) to derive the parameters M_H and α in terms of the inputs $\tan\beta$, M_A and the measured M_h value.

In the case of the charged Higgs boson (whose physics is described by $\tan\beta$, $M_{H^\pm}$ and eventually α), the radiative corrections to $M_{H^\pm}$ are much smaller for large enough M_A and one has, at the few percent level in most cases (which is again smaller than the total $H^\pm$ decay width),

$$M_{H^\pm} \simeq \sqrt{M_A^2 + M_W^2}\,. \tag{16}$$

In conclusion, this approximation allows to ignore the radiative corrections to the Higgs masses and their complicated dependence on the MSSM parameters and to use a simple formula to derive the other parameters of the Higgs sector, α, M_H as well as $M_{H^\pm}$. This considerably simplifies phenomenological analyses in the MSSM Higgs sector which up to now rely either on large scans of the parameter space (as in the previous subsections) or resort to benchmark scenarios in which most of the MSSM parameters are fixed (as is the case of Ref. [70] for instance).

3.2. Implications of the Higgs mass value

We discuss now the implications of the measured mass value of the observed Higgs boson at the LHC [71, 72, 73] that we identify with the lightest state h of the MSSM. We consider the phenomenological MSSM [74] in which the relevant soft SUSY parameters are allowed to vary freely (but with some restrictions) and constrained MSSM scenarios such as the minimal supergravity (mSUGRA) [75], gauge mediated (GMSB) [76] and anomaly mediated (AMSB) [77] supersymmetry breaking models (for a review, see again Ref. [6]). We also discuss the implications of such an M_h value for scenarios in which the supersymmetric spectrum is extremely heavy, the so-called split SUSY [78] or high-scale SUSY models [79].

3.2.1. The phenomenological MSSM.
In an unconstrained MSSM, there is a large number of soft SUSY-breaking parameters, $\mathcal{O}(100)$, but analyses can be performed in the so-called "phenomenological MSSM" (pMSSM) [74], in which CP conservation, flavour diagonal sfermion mass and coupling matrices and universality of the first and second sfermion generations are imposed. The pMSSM involves then 22 free parameters in addition to those of the SM: besides $\tan\beta$ and M_A, these are the higgsino mass μ, the three gaugino masses $M_{1,2,3}$, the diagonal left- and right-handed sfermion mass parameters $m_{\tilde{f}_{L,R}}$ and the trilinear sfermion couplings A_f.

As discussed above, an estimate of the upper bound on M_h can be obtained by including the corrections that involve only the parameters M_S and X_t. However, to be more precise, one could scan the full pMSSM 22 parameter space in order to include the subleading corrections. To do so, one can use RGE programs such as

Suspect [80] which calculate the Higgs and superparticle spectrum in the MSSM including the most up-to-date information [20].

To obtain the value $M_h^{\max}$ with the full radiative corrections, a large scan of the pMSSM parameters in an uncorrelated way was performed [71, 72] in the domains:

$$1 \leq \tan\beta \leq 60\,,\ 50\ \text{GeV} \leq M_A \leq 3\ \text{TeV}\,,$$
$$-9\ \text{TeV} \leq A_t, A_b, A_\tau \leq 9\ \text{TeV}\,,$$
$$50\ \text{GeV} \leq m_{\tilde{f}_L}, m_{\tilde{f}_R}, M_3 \leq 3\ \text{TeV}\,,$$
$$50\ \text{GeV} \leq M_1, M_2, |\mu| \leq 1.5\ \text{TeV}. \tag{17}$$

The results are shown in Figure 13 (top) where the obtained maximal value $M_h^{\max}$ is displayed as a function of the ratio of parameters X_t/M_S. The resulting values are confronted to the mass range $123\ \text{GeV} \leq M_h \leq 127\ \text{GeV}$ when the parametric uncertainties from the SM inputs such as the top quark mass and the theoretical uncertainties in the determination of M_h are included[4].

For $M_S \lesssim 1$ TeV, only the scenarios with X_t/M_S values close to maximal mixing $X_t/M_S \approx \sqrt{6}$ survive. The no-mixing scenario $X_t \approx 0$ is ruled out for $M_S \lesssim 3$ TeV, while the typical mixing scenario, $X_t \approx M_S$, needs large M_S and moderate to large $\tan\beta$ values. From the scan, one obtains a maximum $M_h^{\max}=136$, 126 and 123 GeV with maximal, typical and zero mixing, respectively.

What are the implications for the lightest stop $\tilde{t}_1$ mass? This is illustrated in Figure 13 (bottom) where the contours are shown in the $[M_S, X_t]$ plane in which one obtains $123 < M_h < 127$ GeV from the pMSSM scan; the regions in which $\tan\beta \lesssim 3, 5$ and 60 are highlighted. One sees again that a large part of the parameter space is excluded if the Higgs mass constraint is imposed. In particular, large M_S values, in general corresponding to large $m_{\tilde{t}_1}$ are favored. However, as $M_S = \sqrt{m_{\tilde{t}_1} m_{\tilde{t}_2}}$, the possibility that $m_{\tilde{t}_1}$ is of the order of a few 100 GeV is still allowed, provided that stop mixing (leading to a significant $m_{\tilde{t}_1}, m_{\tilde{t}_2}$ splitting) is large [72].

Masses above 1 TeV for the scalar partners of light quarks and for the gluinos are also required by the direct searches of SUSY particles at the LHC [82], confirming the need of high M_S values. Nevertheless, relatively light stops as well as electroweak sparticles such as sleptons, charginos and neutralinos are still possible allowing for a "natural SUSY" [66] despite of the value $M_h \approx 125$ GeV. Nevertheless, the direct SUSY searches [82] are constraining more and more this natural scenario.

---

[4]This uncertainty is obtained by comparing the outputs of SuSpect and FeynHiggs [81] which use different schemes for the radiative corrections: while the former uses the $\overline{\text{DR}}$ scheme, the latter uses the on-shell scheme; the difference in the obtained M_h amounts to $\approx \pm 2\text{–}3$ GeV in general. To this, one has to add an uncertainty of ± 1 GeV from the top quark mass measurement at the Tevatron, $m_t = 173 \pm 1$ GeV [24] (as discussed previously, it is not entirely clear whether this mass is indeed the pole mass [27]).

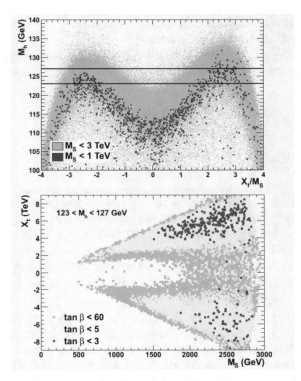

FIGURE 13. The maximal value of the h boson mass as a function of X_t/M_S in the pMSSM when all other soft SUSY-breaking parameters and $\tan\beta$ are scanned (top) and the contours for the Higgs mass range $123 < M_h < 127$ GeV in the $[M_S, X_t]$ plane for some selected $\tan\beta$ values (bottom) [71].

3.2.2. Constrained MSSM scenarios. In constrained MSSM scenarios (cMSSM), the various soft SUSY-breaking parameters obey a number of universal boundary conditions at a high energy scale, thus reducing the number of basic input parameters to a handful. The various soft SUSY-breaking parameters are evolved via the MSSM renormalisation group equations down to the low energy scale M_S where the conditions of proper electroweak symmetry breaking (EWSB) are imposed.

Three classes of such models have been widely discussed in the literature. There is first the minimal supergravity (mSUGRA) model [75] in which SUSY-breaking is assumed to occur in a hidden sector which communicates with the visible sector only via flavour-blind gravitational interactions, leading to universal soft breaking terms, namely a common $m_{1/2}, m_0, A_0$ values for the gaugino masses, sfermion masses and sfermion trilinear couplings. Then come the gauge mediated [76] and anomaly mediated [77] SUSY-breaking (GMSB and AMSB) scenarios in which SUSY-breaking is communicated to the visible sector via, respectively, gauge interactions and a super-Weyl anomaly. These models are described by $\tan\beta$, the sign of μ and a few continuous parameters. Besides of allowing for both signs of μ, requiring $1 \le \tan\beta \le 60$ and, to avoid excessive fine-tuning in the EWSB conditions, imposing the bound $M_S < 3$ TeV, we adopt the following ranges for the inputs of these scenarios:

mSUGRA : $50\,\mathrm{GeV} \le m_0 \le 3\,\mathrm{TeV}$, $50\,\mathrm{GeV} \le m_{1/2} \le 3\mathrm{TeV}$, $|A_0| \le 9\,\mathrm{TeV}$;

GMSB : $10\,\mathrm{TeV} \le \Lambda \le 1000\,\mathrm{TeV}$, $1 \le M_{\mathrm{mess}}/\Lambda \le 10^{11}$, $N_{\mathrm{mess}} = 1$;

AMSB : $1\,\mathrm{TeV} \le m_{3/2} \le 100\,\mathrm{TeV}$, $50\,\mathrm{GeV} \le m_0 \le 3\,\mathrm{TeV}$.

Hence, in contrast to the pMSSM, the various parameters which enter the radiative corrections to M_h are not all independent in these constrained scenarios, as a consequence of the relations between SUSY breaking parameters that are set at the high-energy scale and the requirement that electroweak symmetry breaking is triggered radiatively for each set of input parameters. The additional constraints make that it is not possible to freely tune the parameters that enter the Higgs sector to obtain the pMSSM maximal value of M_h. In order to obtain even a rough determination of M_h^{max} in a given constrained SUSY scenario, it is necessary to scan through the allowed range of values for the basic input parameters.

Using again the program Suspect, a full scan of these scenarios has been performed in Ref. [71] and the results for M_h^{max} are shown in the left-hand side of Figure 14 as a function of $\tan\beta$, the input parameter that is common to all models, and in the right-hand side of the figure as a function of M_S. In the adopted parameter space of the models and with the central values of the SM inputs, the obtained upper h mass value is $M_h^{\mathrm{max}} \approx 121$ GeV in the AMSB scenario, i.e., much less that 125 GeV, while in the GMSB scenario one has $M_h^{\mathrm{max}} \approx 122$ GeV (these values are obtained for $\tan\beta \approx 20$). Thus, clearly, these two scenarios are disfavoured if the lightest h particle has indeed a mass in the range 123–127 GeV and $M_S \lesssim 3$ TeV. In mSUGRA, one obtains $M_h^{\mathrm{max}} = 128$ GeV and, thus, some parameter space would still survive.

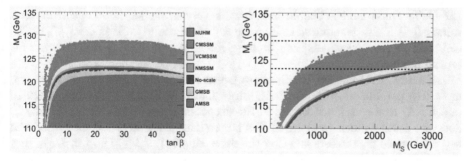

FIGURE 14. The maximal value of the h boson mass as a function of $\tan\beta$ (left) and M_S (right) with a scan of all other parameters in various constrained MSSM scenarios. The range $123 < M_h < 129$ GeV for the light h boson mass is highlighted. From Ref. [71].

The upper bound on M_h in these scenarios can be qualitatively understood by considering in each model the allowed values of the trilinear coupling A_t, which essentially determines the stop mixing parameter X_t and thus the value of M_h for a given scale M_S. In GMSB, one has $A_t \approx 0$ at relatively high scales and its magnitude does not significantly increase in the evolution down to the scale M_S; this implies that we are almost in the no-mixing scenario which gives a low value of M_h^{max} as can be seen from Figure 13. In AMSB, one has a non-zero A_t that is fully predicted at any renormalisation scale in terms of the Yukawa and gauge couplings; however, the ratio A_t/M_S with M_S determined from the overall SUSY breaking scale $m_{3/2}$ turns out to be rather small, implying again that we are close to the no-mixing scenario. Finally, in the mSUGRA model, since we have allowed A_t to vary in a wide range as $|A_0| \leq 9$ TeV, one can get a large A_t/M_S ratio which leads to a heavier Higgs. However, one cannot easily reach the maximal mixing scenario values $X_t/M_S \approx \sqrt{6}$ so that the higher upper bound on M_h in the pMSSM cannot be reached.

In the case of mSUGRA, one can study several interesting special cases: the no-scale scenario with $m_0 \approx A_0 \approx 0$ [83], the scenario $m_0 \approx 0$ and $A_0 \approx -\frac{1}{4}m_{1/2}$ which approximately corresponds to the constrained next-to-MSSM (cNMSSM) [84], $A_0 \approx -m_0$ which corresponds to a very constrained MSSM (VCMSSM) [85], and a non-universal Higgs mass model (NUHM) [86] in which the soft SUSY-breaking scalar mass terms are different for the sfermions and the Higgs doublet fields.

In two particular cases, namely the "no-scale" and the "approximate cN-MSSM" scenarios, the upper bound on M_h is much lower than in the more general mSUGRA case and, in fact, barely reaches $M_h \approx 123$ GeV. The main reason is that these scenarios involve small values of A_0 at the GUT scale, $A_0 \approx 0$ for no-scale and $A_0 \approx -\frac{1}{4}m_{1/2}$ for the cNMSSM which lead to A_t values at the weak scale that are too low to generate a significant stop mixing and, hence, one is again close to the no-mixing scenario. Thus, only a very small fraction of the parameter space of these two sub-classes of the mSUGRA model survive if we impose $123 < M_h < 127$ GeV. These models should thus have a very heavy sfermion spectrum as a value $M_S \gtrsim 3$ TeV is required to increase M_h^{max}. In the VCMSSM case, $M_h \simeq 125$ GeV can be reached as $|A_0|$ can be large for large m_0, $A_0 \approx -m_0$, allowing for typical mixing.

Finally, since the NUHM is more general than mSUGRA as we have two more free parameters, the $[\tan\beta, M_h]$ area shown in Figure 14 is larger than in mSUGRA. However, since we are in the decoupling regime and the value of M_A does not matter much (as long as it a larger than a few hundred GeV) and the key weak-scale parameters entering the determination of M_h, i.e., $\tan\beta, M_S$ and A_t are approximately the same in both models, one obtains a bound M_h^{max} that is only slightly higher in NUHM compared to the mSUGRA case.

In these constrained scenarios and, in particular in the general mSUGRA model, most of the scanned points giving the appropriate Higgs mass correspond to the decoupling regime of the MSSM Higgs sector and, hence, to an h boson with a SM-Higgs cross section and branching ratios. Furthermore, as the resulting SUSY spectrum for $M_h = 125 \pm 2$ GeV is rather heavy in these scenarios (easily evading the LHC limits from direct sparticle searches [82]), one obtains very small contributions to observables like the anomalous muon magnetic moment $(g-2)_\mu$ and to B-physics observables such as the rates BR$(B_s \to \mu^+ \mu^-)$ or BR$(b \to s\gamma)$ [87]. Hence, the resulting spectrum complies with all currently available constraints. In addition, as will be discussed later, the correct cosmological density for the LSP neutralino required by recent measurements [54] can be easily satisfied. The M_h value provides thus a unique constraint in this decoupling regime.

3.2.3. Split and high-scale SUSY models. In the preceding discussion, we have always assumed that the SUSY-breaking scale is relatively low, $M_S \lesssim 3$ TeV, which implies a natural SUSY scenario [66] with supersymmetric and heavier Higgs particles that could be observed at the LHC. However, this choice is mainly dictated by fine-tuning considerations which are a rather subjective matter as there is no compelling criterion to quantify the acceptable amount of tuning. One could well abandon the SUSY solution to the hierarchy problem and have a very high M_S which implies that, except for the lightest h boson, no other scalar particle is accessible at the LHC or at any foreseen collider.

This argument has been advocated to construct the so-called split SUSY scenario [78] in which the soft SUSY-breaking mass terms for all the scalars of the theory, except for one Higgs doublet, are extremely large, i.e., their common value M_S is such that $M_S \gg 1$ TeV (such a situation occurs, e.g., in some string motivated models [88].). Instead, the mass parameters for the spin-$\frac{1}{2}$ particles, the gauginos and the higgsinos, are left in the vicinity of the EWSB scale, allowing for a solution to the dark matter problem and a successful gauge coupling unification, the two other SUSY virtues. The split SUSY models are much more predictive than the usual pMSSM as only a handful parameters are needed to describe the low energy theory. Besides the common value M_S of the soft SUSY-breaking sfermion and one Higgs mass parameters, the basic inputs are essentially the three gaugino masses $M_{1,2,3}$ (which can be unified to a common value at $M_{\rm GUT}$ as in mSUGRA), the higgsino parameter μ and $\tan\beta$. The trilinear couplings A_f, which are expected to have values close to the EWSB scale set by the gaugino/higgsino masses that are much smaller than M_S, will play a negligible role.

Concerning the Higgs sector, the main feature of split SUSY is that at the scale M_S, the boundary condition on the quartic Higgs coupling is determined by SUSY:

$$\lambda(M_S) = \frac{1}{4} \left[g^2(M_S) + g'^2(M_S) \right] \cos^2 2\beta \tag{18}$$

where g and g' are the SU(2) and U(1) gauge couplings. Here, $\tan\beta$ is not a parameter of the low-energy effective theory as it enters only the boundary condition above and cannot be interpreted as the ratio of the two Higgs vevs.

If the scalars are very heavy, they will lead to radiative corrections in the Higgs sector that are significantly enhanced by large logarithms, $\log(M_S/M_{\rm EWSB})$

where $M_{\text{EWSB}} \approx |\mu|, M_2$. In order to have reliable predictions, one has to properly decouple the heavy states from the low-energy theory and resum the large logarithmic corrections; in addition, the radiative corrections due to the gauginos and the higgsinos have to be implemented. Following the early work of Ref. [78], a comprehensive study of the split SUSY spectrum has been performed in Ref. [89]. All the features of the model have been implemented in the code SuSpect [80] upon which the analysis presented in Ref. [71] and summarised here is based.

One can adopt an even more radical attitude than in split SUSY and assume that the gauginos and higgsinos are also very heavy, with a mass close to the scale M_S; this is the case in the so-called high-scale SUSY model [79]. Here, one abandons not only the SUSY solution to the fine-tuning problem but also the solution to the dark matter problem by means of the LSP and the successful unification of the gauge couplings. However, there will still be a trace of SUSY at low energy: the matching of the SUSY and low-energy theories is indeed encoded in the Higgs quartic coupling λ of eq. (18). Hence, even if broken at very high scales, SUSY would still lead to a "light" Higgs whose mass will give information on M_S and $\tan\beta$.

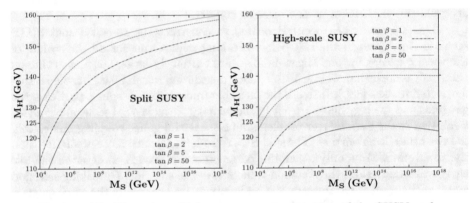

FIGURE 15. The value of h boson mass as a function of the SUSY scale M_S for several values of $\tan\beta = 1, 2, 5, 50$ in the split-SUSY (left) and high-scale SUSY (right) scenarios. From Ref. [71].

The treatment of the Higgs sector of the high-scale SUSY scenario is similar to that of split SUSY: one simply needs to decouple the gauginos and higgsinos from the low energy spectrum (in particular remove their contributions to the renormalisation group evolution of the gauge and Yukawa couplings and to the radiative corrections to M_h) and set their masses to M_S. The version of the program Suspect which handles the split SUSY case can be adapted to also cover the $M_1 \approx M_2 \approx M_3 \approx |\mu| \approx M_S$ case.

Using this tool, a scan in the $[\tan\beta, M_S]$ plane has been performed to determine the value of M_h in the split SUSY and high-scale SUSY scenarios; in the former case, $M_{\text{EWSB}} \approx \sqrt{|M_2\mu|} \approx 246$ GeV was chosen for the low scale. The results are shown in Figure 15 where M_h is displayed as a function of M_S for selected values of $\tan\beta$ in both split (left plot) and high-scale (right plot) SUSY.

As expected, the maximal M_h values are obtained at high $\tan\beta$ and M_S values and, at the scale $M_S \approx 10^{16}$ GeV at which the couplings g and g' approximately unify in the split SUSY scenario, one obtains $M_h \approx 160$ GeV for the higher $\tan\beta = 50$ value. Not included are the error bands in the SM inputs that would lead to an uncertainty of about 2 GeV on M_h, which is now mainly due to the 1 GeV uncertainty on m_t. In addition, the zero-mixing scenario was assumed as the parameter A_t is expected to be much smaller than M_S; this approximation might not be valid for M_S values below 10 TeV and a maximal mixing $A_t/M_S = \sqrt{6}$ would increase the Higgs mass value by up to 10 GeV at $M_S = \mathcal{O}(1 \text{ TeV})$ as was discussed earlier for the pMSSM. In the high-scale SUSY scenario, one obtains a value $M_h \approx 142$ GeV (with again an uncertainty of approximately 2 GeV from the top mass) for high $\tan\beta$ values and at the unification scale $M_S \approx 10^{14}$ GeV [79]. Much smaller M_h values, in the 120 GeV range, can be obtained for lower scales and $\tan\beta$.

Hence, the requirement that the Higgs mass is in the range $123 \lesssim M_h \lesssim 127$ GeV imposes strong constraints on the parameters of these two models. For this mass range, very large scales are needed for $\tan\beta \approx 1$ in the high-scale SUSY scenario, while scales not too far from $M_S \approx 10^4$ GeV are required at $\tan\beta \gg 1$ in both the split and high-scale scenarios. In this case, SUSY should manifest itself at scales much below M_{GUT} if $M_h \approx 125$ GeV.

3.2.4. Splitting the Higgs and sfermion sectors. In the previous high scale scenarios, the Higgs mass parameters were assumed to be related to the mass scale of the scalar fermions in such a way that the masses of the heavier Higgs particles are also of the order of the SUSY scale, $M_A \approx M_S$. However, this needs not to be true in general and one can, for instance, have a NUHM-like scenario where the Higgs masses are decoupled from those of the sfermions. If one is primarily concerned with the MSSM Higgs sector, one may be rather conservative and allow any value for M_A irrespective of the SUSY-breaking scale M_S. This is the quite "model-independent" approach that has been advocated in Refs. [68, 90]: take M_A as a free parameter of the pMSSM, with values ranging from $\mathcal{O}(100 \text{ GeV})$ up to $\mathcal{O}(M_S)$, but make no restriction on M_S which can be set to any value.

An important consequence of this possibility is that it reopens the low $\tan\beta$ region, $\tan\beta \lesssim 3$, that was long thought to be forbidden if one requires a SUSY scale $M_S \lesssim 1$ TeV, as a result of the limit $M_h \gtrsim 114$ GeV from the negative search of a SM-like Higgs boson at LEP [8]. If the SUSY scale is large enough, these small $\tan\beta$ values would become viable again. To estimate the required magnitude of M_S, one can still use Suspect in which the possibility $M_S \gg 1$ TeV is implemented [89] with the full set of radiative corrections up to two-loops included. In Figure 16, displayed are the contours in the plane $[\tan\beta, M_S]$ for fixed mass values $M_h = 120$–132 GeV of the observed Higgs state (these include a 3 GeV theoretical uncertainty and also a 3 GeV uncertainty on the top quark mass [27] that is conservatively

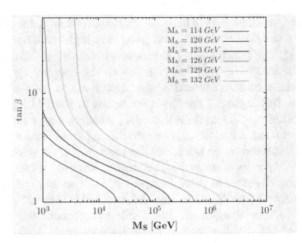

FIGURE 16. Contours for fixed values $M_h = 120, 123, 126, 129$ and 132 GeV in the $[\tan\beta, M_S]$ plane in the decoupling limit $M_A \gg M_Z$; the "LEP2 contour" for $M_h = 114$ GeV is also shown.

added linearly in the extreme cases). The maximal mixing $X_t = \sqrt{6}M_S$ scenario is assumed with 1 TeV gaugino/higgsino mass parameters.

One observes that values of $\tan\beta \approx 1$ are possible and allow for an acceptable M_h provided the scale M_S is large enough. For instance, while one can accommodate a scale $M_S \approx 1$ TeV with $\tan\beta \approx 5$, a large scale $M_S \approx 20$ TeV is required to obtain $\tan\beta \approx 2$; to reach the limit $\tan\beta = 1$ needs an order of magnitude increase of M_S. Outside the decoupling regime, the obtained M_S for a given M_h will be of course larger. For completeness, also shown is the contour for the LEP2 limit $M_h = 114$ GeV which illustrates the fact that $\tan\beta \approx 1$ is still allowed provided that $M_S \gtrsim 20$ TeV.

3.3. Implications of the Higgs production rates

In the MSSM, the couplings of the CP-even Higgs particles h and H to gauge bosons and fermions, compared to the SM Higgs couplings, are changed by factors that involve the sine and the cosine of the mixing angles β and α. Outside the decoupling regime where they reach unity, the reduced couplings (i.e., normalized to their SM values) of the lighter h state to third generation t, b, τ fermions and gauge bosons $V = W/Z$ are for instance given by

$$c_V^0 = \sin(\beta - \alpha), c_t^0 = \cos\alpha / \sin\beta, c_b^0 = -\sin\alpha / \cos\beta. \qquad (19)$$

They thus depend not only on the two inputs $[\tan\beta, M_A]$ as it occurs at tree-level but, a priori, on the entire MSSM spectrum as a result of the radiative corrections, in the same way as the Higgs masses. In principle, as discussed earlier, knowing $\tan\beta$ and M_A and fixing M_h to its measured value, the couplings can be determined. However, this is true when only the radiative corrections to the Higgs masses are included. Outside the regime in which the pseudoscalar A boson and the supersymmetric particles are heavy, there are also direct corrections to the Higgs couplings not contained in the mass matrix of eq. (10) and which alter this simple picture.

First, in the case of b-quarks, additional one-loop vertex corrections modify the tree-level $hb\bar{b}$ coupling: they grow as $m_b\mu\tan\beta$ and can be very large at high $\tan\beta$. The dominant component comes from the SUSY-QCD corrections with sbottom-gluino loops that can be approximated by

$$\Delta_b \simeq 2\alpha_s/(3\pi) \times \mu m_{\tilde{g}}\tan\beta/\max(m_{\tilde{g}}^2, m_{\tilde{b}_1}^2, m_{\tilde{b}_2}^2)$$

[91]. Outside the decoupling regime the c_b coupling reads

$$c_b \approx c_b^0 \times [1 - \Delta_b/(1+\Delta_b) \times (1+\cot\alpha\cot\beta)], \qquad (20)$$

with $\tan\alpha \to -1/\tan\beta$ for $M_A \gg M_Z$. A large Δ_b would significantly alter the dominant $h \to b\bar{b}$ partial width and affect the branching fractions of all other decays.

In addition, the $ht\bar{t}$ coupling is derived indirectly from the $gg \to h$ production cross section and the $h \to \gamma\gamma$ decay branching ratio, two processes that are generated by triangular loops. In the MSSM, these loops involve not only the top quark (and the W boson in the decay $h \to \gamma\gamma$) but also contributions from supersymmetric particles, if not too heavy. In the case of $gg \to h$ production, only the contributions of stops is generally important. Including the later and working in the limit $M_h \ll m_t, m_{\tilde{t}_{1,2}}$, the coupling c_t from the ggF process[5] is approximated by [92]

$$c_t \approx c_t^0\left[1 + \frac{m_t^2}{4m_{\tilde{t}_1}^2 m_{\tilde{t}_2}^2}(m_{\tilde{t}_1}^2 + m_{\tilde{t}_2}^2 - X_t^2)\right], \qquad (21)$$

which shows that indeed, $\tilde{t}$ contributions can be very large for light stops and for large stop mixing. In the $h \to \gamma\gamma$ decay rate, because the $t, \tilde{t}$ electric charges are the same, the $ht\bar{t}$ coupling is shifted by the same amount. If one ignores the usually small contributions of the other sparticles, the $ht\bar{t}$ vertex can be simply parametrised by the effective coupling of eq. (21).

We note that the h couplings to τ leptons and c quarks do not receive the direct corrections of eqs. (20) and (21) and one should still have $c_c = c_t^0$ and $c_\tau = c_b^0$. However, using $c_{t,b}$ or $c_{t,b}^0$ in this case has almost no impact in practice as these couplings appear only in the branching ratios for the decays $h \to c\bar{c}$ and $\tau^+\tau^-$ which are small and the direct corrections should not be too large. One can thus, in a first approximation, assume that $c_c = c_t$ and $c_\tau = c_b$. Another caveat is due to the invisible Higgs decays discussed earlier which we assume to be absent.

Hence, because of the direct corrections, the Higgs couplings cannot be described only by β and α as in eq. (19). To characterize the Higgs particle at the LHC, at least the three independent h couplings c_t, c_b and $c_V = c_V^0$ should be

---

[5]In the case of the production process $gg/q\bar{q} \to ht\bar{t}$, it is still c_t^0 which should describe the $ht\bar{t}$ coupling, but the constraints on the h properties from this process are presently very weak.

considered [41]. One can thus use the effective Lagrangian of eq. 9 and take advantage of the three-dimensional fit in the space $[c_t, c_b, c_V]$ discussed previous section and displayed in Figure 5 which led to best-fit values $c_t = 0.89$, $c_b = 1.01$ and $c_V = 1.02$.

In scenarios where the direct corrections in eqs. (20)–(21) are not quantitatively significant (i.e., considering either not too large values of $\mu \tan\beta$ or high sfermion masses), one can use the MSSM relations of eq. (19) to reduce the number of effective parameters down to two. This allows to perform two-parameter fits in the planes $[c_V, c_t]$, $[c_V, c_b]$ and $[c_t, c_b]$. As an example, the fit of the signal strengths and their ratios in the $[c_t, c_b]$ plane is displayed in Figure 17. In this two-dimensional case, the best-fit point is located at $c_t = 0.88$ and $c_b = 0.97$, while $c_V \simeq 1$. Note that although for the best-fit point one has $c_b \lesssim 1$, actually $c_b \gtrsim 1$ in most of the 1σ region.

Using the formulae in eq. (15) for the angle α and using the input $M_h \approx 125$ GeV, one can make a fit in the plane $[\tan\beta, M_A]$. This is shown in Figure 18 where the 68%, 95% and 99%CL contours from the signal strengths and their ratios are displayed when the theory uncertainty is taken as a bias. The best-fit point when the latter uncertainty is set to zero, is obtained for the values $\tan\beta = 1$ and $M_A = 557$ GeV, which implies for the other parameters using $M_h = 125$ GeV : $M_H = 580$ GeV, $M_{H^\pm} = 563$ GeV and $\alpha = -0.837$ rad which leads to $\cos(\beta - \alpha) \simeq -0.05$. Such a point with $\tan\beta \approx 1$ implies an extremely large value of the SUSY scale, $M_S = \mathcal{O}(100)$ TeV, for $M_h \approx 125$ GeV. One should note, however, that the χ^2 value is relatively stable all over the 1σ region. Hence, larger values of $\tan\beta$ (and lower values of M_A) could also be accommodated reasonably well by the fit, allowing thus for not too large M_S values. In all cases, one has $M_A \gtrsim 200$ GeV though.

3.4. Implications from heavy Higgs searches

We turn now to the constraints on the MSSM Higgs sector that can be obtained from the search of the heavier H/A and $H^\pm$ states at the LHC and start with a brief summary of their production and decay properties.

3.4.1. H, A, H$^\pm$ decays and production at the LHC.

The production and decay pattern of the MSSM Higgs bosons crucially depend on $\tan\beta$. In the decoupling regime that is indicated by the h properties, the heavier CP-even H boson has approximately the same mass as the A state and its interactions are similar. Hence, the MSSM Higgs spectrum will consist of a SM-like Higgs $h \equiv H_{\text{SM}}$ and two pseudoscalar-like particles, $\Phi = H/A$. The $H^\pm$ boson will also be mass degenerate with the Φ states and the intensity of its couplings to fermions will be similar. In the high $\tan\beta$ regime, the couplings of the non-SM like Higgs bosons to b quarks and to τ leptons are so strongly enhanced, and the couplings to top quarks and massive gauge bosons suppressed, that the pattern is rather simple.

This is first the case for the decays: the $\Phi \to t\bar{t}$ channel and all other decay modes are suppressed to a level where their branching ratios are negligible and the

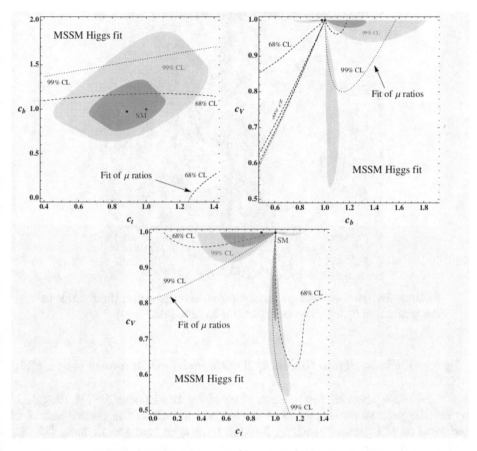

FIGURE 17. Best-fit regions at 68% and 99%CL for the Higgs signal strengths and their ratios in the plane $[c_t, c_b]$, $[c_b, c_V]$ and $[c_t, c_V]$. The best-fit point is indicated in blue. From Ref. [41].

Φ states decay almost exclusively into $\tau^+\tau^-$ and $b\bar{b}$ pairs, with branching ratios of $BR(\Phi \to \tau\tau) \approx 10\%$ and $BR(\Phi \to b\bar{b}) \approx 90\%$. The $H^\pm$ boson decay into $\tau\nu_\tau$ final states with a branching fraction of almost 100% for $H^\pm$ masses below the tb threshold, $M_{H^\pm} \lesssim m_t + m_b$, and a branching ratio of only $\approx 10\%$ for masses above this threshold while the rate for $H^\pm \to tb$ will be at the $\approx 90\%$ level in most cases.

Concerning the production, the strong enhancement of the b-quark couplings at high $\tan\beta$ makes that only two processes are relevant in this case: $gg \to \Phi$ fusion with the b-loop included and associated production with b-quarks, $gg/q\bar{q} \to b\bar{b} + \Phi$, which is equivalent to the fusion process $b\bar{b} \to \Phi$ with no-additional final b-quark. All other processes, in particular $V\Phi$, $t\bar{t}\Phi$ and VBF have suppressed rates. In both the $b\bar{b}$ and gg fusion cases, as $M_\Phi \gg m_b$, chiral symmetry holds and the rates are approximately the same for the CP-even H and CP-odd A bosons. While

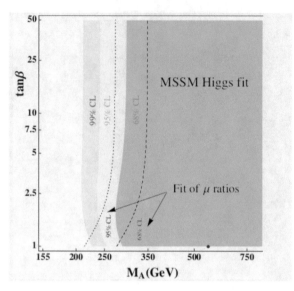

FIGURE 18. Best-fit regions for the signal strengths and their ratios in the plane [$\tan\beta$, M_A]; the best point is in blue [41].

$\sigma(gg \to \Phi)$ is known up to NLO in QCD [93], $\sigma(bb \to \Phi)$ is instead known up to NNLO [94].

The most powerful LHC search channel for the heavier MSSM Higgses is by far the process $pp \to gg + b\bar{b} \to \Phi \to \tau^+\tau^-$ for which the precise values of the cross section times branching fraction have been updated in Refs. [32, 34] and an assessment of the associated theoretical uncertainties has been made. It turns out that, in the production cross section, the total uncertainty from scale variation, the PDFs and α_s as well as from the b-quark mass are not that small: $\Delta^{\mathrm{TH}}\sigma(pp \to \Phi) \times \mathrm{BR}(\Phi \to \tau\tau) \approx \pm 25\%$ in the entire M_Φ range probed at the LHC at $\sqrt{s} = 8$ TeV; Figure 19. Besides the QCD uncertainty, three other features could alter the rate $\sigma(pp \to \Phi \to \tau\tau)$ in the MSSM and they are related to the impact of the SUSY particle contributions:

i) In the case of H (A does not couple to identical sfermions), there are squark (mainly stop) loops that contribute in addition in the $gg \to H$ process. But as they are damped by powers of $\tilde{m}_Q^2$ for $M_H \lesssim 2m_Q^2$, these should be small for $\tilde{m}_Q \gtrsim 1$ TeV, in particular at high $\tan\beta$ where the b-contribution is strongly enhanced.

ii) The vertex correction to the $\Phi b\bar{b}$ couplings, Δ_b of eq. (20), grows as $\mu \tan\beta$ and can be very large in the high $\tan\beta$ regime. However, in the full process $pp \to \Phi \to \tau^+\tau^-$, this correction appears in both the cross section and the branching fraction and largely cancels outs as one obtains, $\sigma \times \mathrm{BR} \times (1 - \Delta_b/5)$. A very large contribution $\Delta_b \approx 1$ changes the rate only by 20%, i.e., less than the QCD uncertainty.

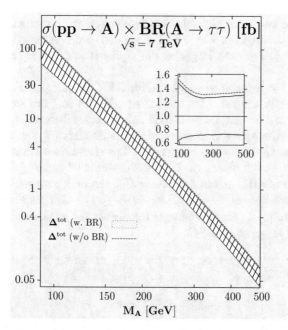

FIGURE 19. The combined $\sigma(pp \to A) \times \mathrm{BR}(A \to \tau\tau)$ rate with theoretical uncertainties with and without the branching ratio; in the inserts, shown are the uncertainties when the rates are normalized to the central values. From Ref. [34].

iii) The possibility of light sparticles would lead to the opening of H/A decays into SUSY channels that would reduce $\mathrm{BR}(\Phi \to \tau\tau)$. For $M_\Phi \lesssim 1$ TeV, the only possibilities are decays into light neutralinos or charginos and sleptons. These are in general disfavored at high $\tan\beta$ as the $\Phi \to b\bar{b} + \tau\tau$ modes are strongly enhanced.

Thus, it is only in the unlikely cases where $\mathrm{BR}(H \to \tilde{\tau}_1 \tilde{\tau}_1)$ is of order 50%, the squark loop contribution to the $gg \to H$ process is of the order 50%, or the Δ_b SUSY correction is larger than 100%, that one can change the $pp \to \Phi \to \tau\tau$ rate by $\approx 25\%$, which is the level of the QCD uncertainty. One thus expects $\sigma(pp \to \Phi) \times \mathrm{BR}(\Phi \to \tau\tau)$ to be extremely robust and to depend almost exclusively on M_A and $\tan\beta$.

Finally, for the charged Higgs boson, the dominant search channel is $H^\pm \to \tau\nu$ with the $H^\pm$ produced in top quark decays for $M_{H^\pm} \lesssim m_t - m_b \approx 170$ GeV, $pp \to t\bar{t}$ with $t \to H^+ b \to \tau\nu b$. This is particularly true at high $\tan\beta$ when $\mathrm{BR}(t \to H^+ b) \propto \bar{m}_b^2 \tan^2\beta$ is significant. For higher $H^\pm$ masses, one should rely on the three-body production process $pp \to tbH^\pm \to tb\tau\nu$ but the rates are presently rather small.

In the low $\tan\beta$ regime, $\tan\beta \lesssim 5$, the phenomenology of the heavier $A, H, H^\pm$ bosons is richer [68, 95]. Starting with the cross sections, we display in Figure 20 the rates for the relevant production processes at the LHC with $\sqrt{s} = 8$ TeV assuming $\tan\beta = 2.5$. For smaller $\tan\beta$ values, the rates except for $pp \to H/A + b\bar{b}$ are even larger as the $H/A+tt$ and HVV couplings are less suppressed.

Because of CP invariance which forbids AVV couplings, there is no AV and Aqq processes while the rates for associated $t\bar{t}A$ and $b\bar{b}A$ are small because the Att (Abb) couplings are suppressed (not sufficiently enhanced) compared to the SM Higgs. Only the $gg \to A$ process with the dominant t and sub-dominant b contributions included provides large rates. The situation is almost the same for H: only $gg \to H$ is significant at $M_H \gtrsim 300$ GeV and $\tan\beta \lesssim 5$; the VBF and HV modes give add little at $\tan\beta \approx 1$. For $H^{\pm}$, the dominant production channel is again top quark decays, $t \to H^+b$ for $M_{H^\pm} \lesssim 170$ GeV as for $\tan\beta \lesssim 5$, the $m_t/\tan\beta$ piece of $g_{H^\pm tb}$ becomes large; for higher $H^{\pm}$ masses, the main process to be considered is $gg/q\bar{q} \to H^{\pm}tb$.

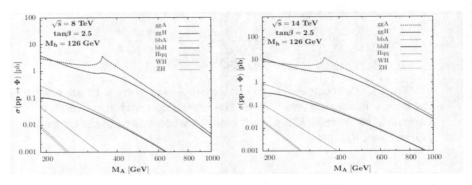

FIGURE 20. The production cross sections of the MSSM heavier neutral Higgs bosons at the LHC at $\sqrt{s} = 8$ for $\tan\beta = 2.5$; only the main production channels are considered [68].

Turning to the $H/A/H^{\pm}$ decay pattern, it can be rather involved at low $\tan\beta$. A summary is as follows for $\tan\beta \lesssim 3$; Figure 21 shows the rates for $\tan\beta = 2.5$. i) Above the $t\bar{t}$ (tb) threshold for $H/A(H^{\pm})$, the decay channels $H/A \to t\bar{t}$ and $H^+ \to t\bar{b}$ are by far dominant for $\tan\beta \lesssim 3$ and do not leave space for any other mode. ii) Below the $t\bar{t}$ threshold, the $H \to WW, ZZ$ decay rates are still significant as g_{HVV} is not completely suppressed. iii) For $2M_h \lesssim M_H \lesssim 2m_t$, $H \to hh$ is the dominant H decay mode as the Hhh self-coupling is large at low $\tan\beta$. iv) For $M_A \gtrsim M_h + M_Z$, $A \to hZ$ decays would occur but the $A \to \tau\tau$ channel is still important with rates $\gtrsim 5\%$. v) In the case of $H^{\pm}$, the channel $H^+ \to Wh$ is important for $M_{H^\pm} \lesssim 250$ GeV, similarly to the $A \to hZ$ case.

3.4.2. Constraints from the LHC Higgs searches. The most efficient channel to probe the heavier MSSM Higgs bosons is by far $pp \to gg+bb \to H/A \to \tau^+\tau^-$. Searches for this process have been performed by ATLAS with ≈ 5 fb^{-1} data at

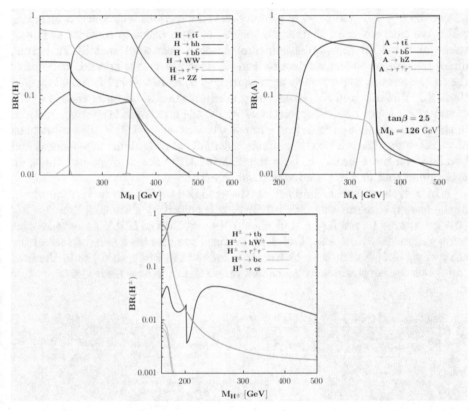

FIGURE 21. The $H/A/H^{\pm}$ branching ratios as functions of the Higgs masses for $\tan\beta = 2.5$ [68].

the 7 TeV run [96] and by CMS with $\approx 5 + 12$ fb^{-1} data at the 7 TeV and 8 TeV runs [97]. Upper limits on the production cross section times decay branching ratio have been set and they can be turned into constraints on the MSSM parameter space.

In the left-hand side of Figure 22 displayed is the sensitivity of the CMS $pp \rightarrow \Phi \rightarrow \tau\tau$ analysis with 17 fb^{-1} of data in the $[\tan\beta, M_A]$ plane. The excluded region, obtained from the observed limit at the 95%CL is drawn in blue. The dotted line represents the median expected limit which turns out to be weaker than the observed limit. As can be seen, this constraint is extremely restrictive and for values $M_A \lesssim 250$ GeV, it excludes almost the entire intermediate and high $\tan\beta$ regimes, $\tan\beta \gtrsim 5$. The constraint is less effective for a heavier A boson, but even for $M_A \approx 400$ GeV the high $\tan\beta \gtrsim 10$ region is excluded and one is even sensitive to large values $M_A \approx 800$ GeV for $\tan\beta \gtrsim 50$.

There are, however, some caveats to this exclusion limit as discussed previously. The first one is that there is a theoretical uncertainty of order of $\pm 25\%$ that affects the $gg \to \Phi$ and $b\bar{b} \to \Phi$ production cross sections which, when included, will make the constraint slightly weaker as one then needs to consider the lower value predicted for the production rate. A second caveat is that SUSY effects, direct corrections to the production and H/A decays into sparticles, could alter the rate. However, as previously argued, $\sigma(pp\to\Phi)\times\mathrm{BR}(\Phi\to\tau\tau)$ is robust against these SUSY effects and the latter will unlikely make a substantial change of the cross section times branching fraction. Finally, the constraint is specifically given in the maximal mixing scenario $X_t/M_S = \sqrt{6}$ with $M_S = 1$ TeV. The robustness of $\sigma\times\mathrm{BR}$ makes that the exclusion limit is actually almost model independent and is valid in far more situations than the "MSSM M_h^{max} scenario" quoted there, an assumption that can be removed without any loss.

In fact, the exclusion limit can also be extended to the low $\tan\beta$ region which, in the chosen scenario with $M_S = 1$ TeV, is excluded by the LEP2 limit on M_h (the green area in the figure) but should resurrect if the SUSY scale is kept as a free parameter. Note also, that H/A bosons have also been searched for in the channel $gg \to b\bar{b}\Phi$ with $\Phi \to b\bar{b}$ (requiring more than 3-tagged b jets in the final state) but the constraints are much less severe than the ones derived from the $\tau\tau$ channel [98].

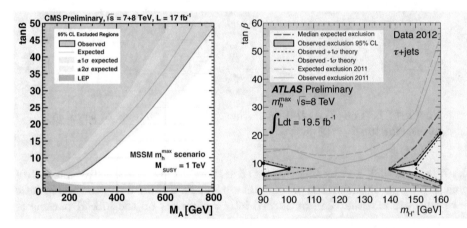

FIGURE 22. Left: the expected and observed exclusion limits in the $[\tan\beta, M_A]$ plane in the CMS search of the MSSM neutral Higgs bosons in the channels $pp \to h/H/A \to \tau^+\tau^-$ with ≈ 17 fb^{-1} data collected at $\sqrt{s} = 7+8$ TeV [97]. Right: the $H^\pm$ limits from ATLAS with $\sqrt{s} = 8$ TeV and ≈ 20 fb^{-1} data in the channel $t \to bH^+ \to b\tau\nu$ [99].

Turning to the H^+ boson [100, 99], the most recent result has been provided by the ATLAS collaboration using the full ≈ 20 fb^{-1} data collected at $\sqrt{s} = 8$ TeV. The $H^\pm$ search as been performed using the τ plus jets channel with a hadronically decaying τ lepton in the final state. For $M_{H^\pm} \lesssim 160$ GeV, the results are shown in Figure 22 (right). Here, the relevant process is top quark decays, $t \to H^+b$ with the decay $H^+ \to \tau\nu$ having a branching ratio of almost 100% at moderate to high $\tan\beta$. For these high values, the H^+tb coupling has a component $\propto m_b \tan\beta$ which makes $\mathrm{BR}(t \to H^+ b)$ rather large. Almost the entire $\tan\beta \gtrsim 10$ region is excluded.

In addition, the branching fraction for the decay $t \to bH^+$ is also large at low $\tan\beta$ when the component of the coupling $g_{tbH^+} \propto \bar{m}_t/\tan\beta$ becomes dominant. On the other hand, the branching fraction for $H^\pm \to \tau\nu$ does not become very small as it has competition only from $H^+ \to c\bar{s}$ which, even for $\tan\beta \approx 1$, does not dominate. Hence, the rates for $pp \to t\bar{t}$ with $t \to bH^+ \to b\tau\nu$ are comparable for $\tan\beta \approx 3$ and $\tan\beta \approx 30$ and the processes can also probe the low $\tan\beta$ region. This is exemplified in Figure 22 (right) where one can see that the entire area below $\tan\beta \approx 5$ is also excluded. Remains then, for $H^\pm$ masses close to 90 GeV (where the detection efficiency is lower) and 160 GeV (that is limited by phase-space), the intermediate $\tan\beta \approx 5\text{--}10$ area where the $H^\pm tb$ coupling is not strongly enhanced

This ATLAS search has been extended to larger values of $M_{H^\pm}$ where the charged Higgs is produced in association with top quarks, $gb \to tH^+$, but the constraints are poor (only the region $\tan\beta \gtrsim 50$ is excluded for $M_{H^\pm} = 200\text{--}300$ GeV) as the cross section for this process is low.

The reopening of the low $\tan\beta$ region allows to consider a plethora of very interesting channels for the heavier Higgs bosons to be also investigated at the LHC: heavier CP-even H decays into massive gauge bosons $H \to WW, ZZ$ and Higgs bosons $H \to hh$, CP-odd Higgs decays into a vector and a Higgs boson, $A \to hZ$, CP-even and CP-odd Higgs decays into top quarks, $H/A \to t\bar{t}$, and even the charged Higgs decay $H^\pm \to Wh$. These final states have been searched for in the context of a heavy SM Higgs boson or for new resonances in some non-SUSY beyond the SM scenarios and the analyses can be adapted to the case of the heavier MSSM Higgs bosons. They would then allow to cover a larger part of the parameter space of the MSSM Higgs sector in a model-independent way, i.e., without using the information on the scale M_S and more generally on the SUSY particle spectrum that appear in the radiative corrections.

In Ref. [68] a preliminary analysis of these channels has been performed using current information given by the ATLAS and CMS collaborations in the context of searches for the SM Higgs boson or other heavy resonances (in particular new Z' or Kaluza–Klein gauge bosons that decay into $t\bar{t}$ pairs). The results are shown in Figure 23 with an extrapolation to the full 25 fb^{-1} data of the 7+8 TeV LHC run (it has been assumed that the sensitivity scales simply as the square root of the number of events). The sensitivities from the usual $H/A \to \tau^+\tau^-$ and $t \to bH^+ \to b\tau\nu$ channels are also shown. The green and red areas correspond to the domains where the $H \to VV$ and $H/A \to t\bar{t}$ channels become constraining. The sensitivities in the $H \to hh$ and $A \to hZ$ modes are given by, respectively, the yellow and brown areas which peak in the mass range $M_A = 250\text{--}350$ GeV that is visible at low $\tan\beta$ values.

The outcome is impressive. These channels, in particular $H \to VV$ and $H/A \to t\bar{t}$, are very constraining as they cover the entire low $\tan\beta$ area that was

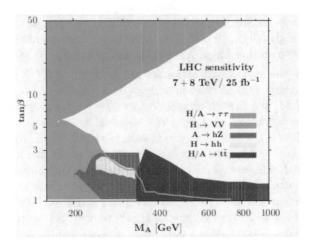

FIGURE 23. The estimated sensitivities in the various search channels for the heavier MSSM Higgs bosons in the [$\tan\beta$, M_A] plane: $H/A \to \tau\tau$, $H \to WW + ZZ$, $H/A \to t\bar{t}$, $A \to hZ$ and $H \to hh$ [68]. The projection is made for the LHC with 7+8 TeV and the full 25 fb^{-1} of data collected so far. The radiative corrections are such that the h mass is $M_h = 126$ GeV.

previously excluded by the LEP2 bound up to $M_A \approx 500$ GeV. Even $A \to hZ$ and $H \to hh$ would be visible at the current LHC in small portions of the parameter space.

4. What next?

The last three years were extremely rich and exciting for particle physics. With the historical discovery of a Higgs boson by the LHC collaborations ATLAS and CMS, crowned by a Nobel price last fall, and the first probe of its basic properties, they witnessed a giant step in the unraveling of the mechanism that breaks the electroweak symmetry and generates the fundamental particle masses. They promoted the SM as the appropriate theory, up to at least the Fermi energy scale, to describe three of Nature's interactions, the electromagnetic, weak and strong forces,

However, it is clear that these two years have also led to some frustration as no signal of physics beyond the SM has emerged from the LHC data. The hope of observing some signs of the new physics models that were put forward to address the hierarchy problem, that is deeply rooted in the Higgs mechanisms, with Supersymmetric theories being the most attractive ones, did not materialize.

The discovery of the Higgs boson and the non-observation of new particles has nevertheless far reaching consequences for supersymmetric theories and, in particular, for their simplest low energy formulation, the MSSM. The mass of approximately 125 GeV of the observed Higgs boson implies that the scale of SUSY-breaking is rather high, at least $\mathcal{O}(\text{TeV})$. This is backed up by the limits on the masses of strongly interacting SUSY particles set by the ATLAS and CMS searches, which in most cases exceed the TeV range [82]. This implies that if SUSY is indeed behind the stabilization the Higgs mass against very high scales that enter via quantum corrections, it is either fine-tuned at the permille level at least or its low energy manifestation is more complicated than expected.

The production and decay rates of the observed Higgs particles, as well as its spin and parity quantum numbers, as measured by the ATLAS and CMS collaborations with the ≈ 25 fb^{-1} data collected at $\sqrt{s}=7$ and 8 TeV, indicate that its couplings to fermions and gauge bosons are approximately SM-like. In the context of the MSSM, this implies that we seem to be in the decoupling regime and this 125 GeV particle can be only identified with the lightest h boson, while the other $H/A/H^\pm$ states must be heavier than approximately the Fermi scale. This last feature is also backed up by LHC direct searches of these heavier Higgs states.

This drives up to the question that is now very often asked in particle physics (and elsewhere): what to do next? The answer is, for me, obvious: we are only in the beginning of a new era[6]. Indeed, it was expected since a long time that the probing of the EWSB mechanism will be at least a two chapters story. The first one is the search and the observation of a Higgs-like particle that will confirm the scenario of the SM and most of its extensions, that is, a spontaneous symmetry breaking by a scalar field that develops a non-zero vacuum expectation value. This long chapter has just been closed by the ATLAS and CMS collaborations with the spectacular observation of a Higgs boson. This observation opens a second and equally important chapter: the precise determination of the Higgs profile and the unraveling of the EWSB mechanism itself.

A more accurate measurement of the Higgs couplings to fermions and gauge bosons will be mandatory to establish the exact nature of the mechanism and, eventually, to pin down effects of new physics if additional ingredients beyond those of the SM are involved. This is particularly true in weakly interacting theories such as SUSY in which the quantum effects are expected to be small. These measurements could be performed at the upgraded LHC with an energy close to $\sqrt{s}=14$ TeV, in particular if a very high luminosity, a few ab^{-1}, is achieved [101, 102].

At this upgrade, besides improving the measurements performed so far, rare but important channels such as associated Higgs production with top quarks, $pp \to t\bar{t}h$, and Higgs decays into $\mu^+\mu^-$ and $Z\gamma$ states could be probed. Above all, a determination of the self-Higgs coupling could be made by searching for double Higgs production, e.g., in the gluon fusion channel $gg \to hh$ [103]; this would be a first step towards the reconstruction of the scalar potential that is responsible of

---

[6]One can rightfully use here the words of Winston Churchill in November 1942 after the battle of El Alamein (which in Arabic literally means "the two flags" but could also mean "the two worlds" or even "the two scientists"!): "Now, this is not the end; it is not even the beginning to the end; but it is, perhaps, the end of the beginning".

EWSB. A proton collider with an energy $\sqrt{s} = 30$ to 100 TeV could do a similar job [102].

In a less near future, a high-energy lepton collider, which is nowadays discussed in various options (ILC, TLEP, CLIC, μ-collider) would lead to a more accurate probing of the Higgs properties [104], promoting the scalar sector to the high-precision level of the gauge and fermion sectors achieved by LEP and SLC [7].

Besides the high precision study of the already observed Higgs, one should also continue to search for the heavy states that are predicted by SUSY, not only the superparticles but also the heavier Higgs bosons. The energy upgrade to ≈ 14 TeV (and eventually beyond) and the planned order of magnitude (or more) increase in luminosity will allow to probe much higher mass scales than presently.

In conclusion, it is not yet time to give up on SUSY and on New Physics in general but, rather, to work harder to be fully prepared for the more precise and larger data that will be delivered by the upgraded LHC. It will be soon enough to "philosophize" then as the physics landscape will become more clearer.

Acknowledgement

This work is supported by the ERC Advanced Grant Higgs@LHC.

References

[1] The ATLAS collaboration. Phys. Lett. **B716**, 1 (2012).

[2] The CMS collaboration. Phys. Lett. **B716**, 30 (2012).

[3] Englert, F., and Brout, R.: Phys. Rev. Lett. **13**, 321 (1964); Higgs, P.: Phys. Lett. **12**, 132 (1964); Higgs, P.: Phys. Rev. Lett. **13**, 508 (1964); Guralnik, G., Hagen, C., and Kibble, T.: Phys. Rev. Lett. **13**, 585 (1964).

[4] Gunion, J., Haber, H., Kane, G., and Dawson, S.: *The Higgs Hunter's Guide*. Reading 1990.

[5] Djouadi, A.: Phys. Rep. **457**, 1 (2008).

[6] Djouadi, A.: Phys. Rep. **459**, 1 (2008).

[7] Olive, K., et al.: Particle Data Group. Chin. Phys. **C38**, 090001 (2014).

[8] LEP collaborations: Phys. Lett. **B565**, 61 (2003).

[9] The LEP collaborations and the LEP electroweak Working Group: hep-ex/0412015; http://lepewwg.web.cern.ch/LEPEWWG.

[10] Baak, M., et al.: the GFITTER collaboration: arXiv:1107.0975 [hep-ph].

[11] Lee, B.W., Quigg, C., and Thacker, H.B.: Phys. Rev. **D16**, 1519 (1977).

[12] Llewellyn Smith, C.H.: Phys. Lett. **B46**, 233 (1973); Bell, J.S.: Nucl. Phys. **B60**, 427 (1973); Cornwall, J., et al.: Phys. Rev. **D10**, 1145 (1974).

[13] Cabibbo, N., Maiani, L., Parisi, G., and Petronzio, R.: Nucl. Phys. **B158**, 295 (1979); Lindner, M., Sher, M., and Zaglauer, H.W.: Phys. Lett. **B228**, 139 (1989); Hambye, T., and Riesselmann, K.: Phys. Rev. **D55**, 7255 (1997); Ellis, J., et al.: Phys. Lett. **B679**, 369 (2009).

[14] Degrassi, G., et al.: arXiv:1205.6497; Bezrukov, F., Kalmykov, M., Kniehl, B., and Shaposhnikov, M.: arXiv:1205.2893.

[15] Drees, M., Godbole, R., and Roy, P.: *Theory and phenomenology of sparticles.* World Sci., 2005; Baer, H., and Tata, X.: *Weak scale Supersymmetry: from superfields to scattering events.* Cambridge U. Press, 2006; S. Martin, hep-ph/9709356.

[16] Okada, Y., Yamaguchi, M., and Yanagida, T.: Prog. Theor. Phys. **85**, 1 (1991); Ellis, J., Ridolfi, G., and Zwirner, F.: Phys. Lett. **B257**, 83 (1991); Haber, H.E., and Hempfling, R.: Phys. Rev. Lett. **66**, 1815 (1991).

[17] Carena, M., Espinosa, J.R., Quiros, M., and Wagner, C.E.: Phys. Lett. **B355**, 209 (1995); Haber, H., Hempfling, R., and Hoang, A.: Z. Phys. **C75**, 539 (1997).

[18] Heinemeyer, S., Hollik, W., and Weiglein, G.: Phys. Rev. **D58**, 091701 (1998); Eur. Phys. J. **C9**, 343 (1999); Degrassi, G., Slavich, P., and Zwirner, F.: Nucl. Phys. **B611**, 403 (2001); Brignole, A., Degrassi, G., Slavich, P., and Zwirner, F.: Nucl. Phys. **B631**, 195 (2002); Nucl. Phys. **B643**, 79 (2002).

[19] Martin, S.: Phys. Rev. **D75**, 055005 (2007); Kant, P., Harlander, R., Mihaila, L., and Steinhauser, M.: JHEP **1008**, 104 (2010); Feng, J., et al.: Phys. Rev. Lett. **111**, 131802 (2013).

[20] Allanach, B., et al.: JHEP **0409**, 044 (2004).

[21] Heinemeyer, S., Hollik, W., and Weiglein, G.: Phys. Rep. **425**, 265 (2006); Heinemeyer, S.: IJMPA **21**, 2659 (2006).

[22] Carena, M., Haber, H.: Prog. Part. Nucl. Phys. **50**, 63 (2003).

[23] See for instance Haber, H.E.: hep-ph/9505240.

[24] The Tevatron, CDF and D0 EW working group: arXiv:1305.3929 [hep-ex].

[25] The ATLAS collaboration: note ATLAS-CONF-2013-014.

[26] The CMS collaboration: note CMS-PAS-HIG-13-005.

[27] Alekhin, S., Djouadi, A., and Moch, S.: Phys. Lett. **B716**, 214 (2012).

[28] Martin, A., et al.: Eur. Phys. J. **C63**, 189 (2009).

[29] Djouadi, A.: Eur. Phys. J. **C73**, 2498 (2013).

[30] The ATLAS collaboration: ATLAS-CONF-2013-034.

[31] The CMS collaboration: CMS-HIG-13-005.

[32] Dittmaier, S., et al.: (LHC Higgs Working Group). arXiv:1101.0593.

[33] Baglio, J., and Djouadi, A.: JHEP **1010**, 064 (2010).

[34] Baglio, J., and Djouadi, A.: JHEP **1103**, 055 (2011).

[35] Dittmaier, S., et al.: (LHC Higgs WG). arXiv:1201.3084.

[36] Zeppenfeld, D., Kinnunen, R., Nikitenko, A., and Richter-Was, E.: Phys. Rev. **D62**, 013009 (2000); Djouadi, A., et al.: hep-ph/0002258; Dührssen, M., et al.: Phys. Rev. **D70**, 113009 (2004); Assamagan, K., et al.: hep-ph/0406152.

[37] Djouadi, A., Kalinowski, J., and Spira, M.: Comput. Phys. Commun. **108**, 56 (1998); Djouadi, A., Muhlleitner, M., and Spira, M.: Acta. Phys. Polon. **B38**, 635 (2007).

[38] Carmi, D., Falkowski, A., Kuflik, E., and Volansky, T.: JHEP **1207**, 136 (2012); Azatov, A., Contino, R., and Galloway, J.: JHEP **1204**, 127 (2012); Espinosa, J., Grojean, C., Muhlleitner, M., and Trott, M.: JHEP **1205**, 097 (2012); JHEP **1212**, 045 (2012); JHEP **1209**, 126 (2012); Giardino, P., Kannike, K., Raidal, M., and

Strumia, A.: JHEP **1206**, 117 (2012); Phys. Lett. **B718**, 469 (2012); Li, T., Wan, X., Wang, Y.-K., and Zhu, S.-H.: JHEP **1209**, 086 (2012); Ellis, J., and You, T.: JHEP **1206**, 140 (2012); Azatov, A., et al.: JHEP **1206**, 134 (2012); Klute, M., Lafaye, R., Plehn, T., Rauch, M., and Zerwas, D.: Phys. Rev. Lett. **109**, 101801 (2012); Chang, J., Cheung, K., Tseng, P.-Y., and Yuan, T.-C.: JHEP **1212**, 058 (2012); Chang, S., Newby, C.A., Raj, N., and Wanotayaroj, C.: Phys. Rev. **D86**, 095015 (2012); Low, I., Lykken, J., and Shaughnessy, G.: Phys. Rev. **D86**, 093012 (2012); Montull, M., and Riva, F.: JHEP **1211**, 018 (2012); Carmi, D., et al.: JHEP **1210**, 196 (2012); Banerjee, S., Mukhopadhyay, S., and Mukhopadhyaya, B.: JHEP **1210**, 062 (2012); Bonnet, F., Ota, T., Rauch, M., and Winter, W.: Phys. Rev. **D86**, 093014 (2012); Plehn, T., and Rauch, M.: Europhys. Lett. **100**, 11002 (2012); Baglio, J., et al.: Phys. Lett. **B716**, 203 (2012); Belanger, G., Dumont, B., Ellwanger, U., Gunion, J., and Kraml, S.: JHEP **1302**, 053 (2013); and arXiv:1306.2941; Altarelli, G.: arXiv:1308.0545; Corbett, T., et al.: Phys. Rev. **D86**, 075013 (2012); Alves, A., et al.: Eur. Phys. J. **C73**, 2288 (2013); Peskin, M.: arXiv:1207.2516v1 [hep-ph]; Cacciapaglia, G., Deandrea, A., La Rochelle, G.D., and Flament, J.B.: JHEP **1303**, 029 (2013); Falkowski, A., Riva, F., and Urbano, A.: arXiv:1303.1812 [hep-ph]. Cheung, C., McDermott, S., and Zurek, K.: JHEP **1304**, 074 (2013); Cheung, K., Lee, J.S., and Tseng, P.: JHEP **1305**, 134 (2013).

[39] Arbey, A., et al.: Phys. Lett. **B720**, 153 (2013).

[40] Djouadi, A., and Moreau, G.: Eur. Phys. J. **C73**, 2512 (2013).

[41] Djouadi, A., Maiani, L., Moreau, G., Polosa, A., Quevillon, J., and Riquer, V.: Eur. Phys. J. **C73**, 2650 (2013), arXiv:1307.5205 [hep-ph].

[42] Djouadi, A., and Lenz, A.: Phys. Lett. **B715**, 310 (2012);

[43] Denner, A., et al.: Eur. Phys. J. **C72**, 1992 (2012); Kuflik, E., Nir, Y., and Volansky, T.: Phys. Rev. Lett. **110**, 091801 (2013); Passarino, G., Sturm, C., and Uccirati, S.: Phys. Lett. **B706**, 195 (2011); Djouadi, A., Gambino, P., Kniehl, B.A.: Nucl. Phys. **B523**, 17 (1998); Djouadi, A., and Gambino, P.: Phys. Rev. Lett. **73**, 2528 (1994).

[44] The CMS collaboration: Phys. Rev. **D86**, 112003 (2012); the ATLAS collaboration: Phys. Rev. Lett. **109**, 032001 (2012).

[45] Kauer, N., and Passarino, G.: JHEP **1208**, 116 (2012).

[46] Caola, F., and Melnikov, K.: Phys. Rev. **D88**, 054024 (2013); Campbell, J., Ellis, R., and Williams, C.: JHEP **1404**, 60 (2014) and Phys. Rev. **D89**, 053011 (2014).

[47] ATLAS collaboration: ATLAS-CONF-2014-042; Khachatryan, V., et al. (CMS Collaboration): Phys. Lett. **B736**, 64 (2014).

[48] Englert, C., and Spannowsky, M.: Phys. Rev. **D90**, 053003 (2014).

[49] The ATLAS collaboration: ATLAS-CONF-2013-011.

[50] The CMS collaboration: CMS-PAS-HIG-13-013.

[51] Djouadi, A., Falkowski, A., Mambrini, Y., and Quevillon, J.: Eur. Phys. J. **C73**, 2455 (2013).

[52] Bai, Y., Draper, P., and Shelton, J.: JHEP **1207**, 192 (2012); Englert, C., Jaeckel, J., Re, E., and Spannowsky, M.: Phys. Rev. **D85**, 035008 (2012).

[53] The CMS collaboration: arXiv:1206.5663 [hep-ex]; The ATLAS collaboration: ATLAS-CONF-2012-147.

[54] For a recent review see, e.g., Catena, R., and Covi, L.: arXiv:1310.4776.

[55] Djouadi, A., Lebedev, O., Mambrini, Y., Quevillon, J.: Phys. Lett. **B709**, 65 (2012).

[56] Landau, L.: Dokl. Akad. Nauk Ser. Fiz. **60**, 207 (1948); Yang, C.: Phys. Rev. **77**, 242 (1950).

[57] Ellis, J., Sanz, V., and You, T.: Phys. Lett. **B726**, 244 (2013).

[58] For a review of the CP Higgs issue, see, e.g., Kraml, S., (ed.) et al., hep-ph/0608079; a more recent study with references, see A. Alloul, B. Fuks and V. Sanz, arXiv:1310.5150.

[59] The ATLAS collaboration: Phys. Lett. **B726**, 120 (2013); the CMS collaboration: Phys. Rev. Lett. **110**, 081803 (2013).

[60] Plehn, T., Rainwater, D., and Zeppenfeld, D.: Phys. Rev. Lett. **88**, 051801 (2002).

[61] Hagiwara, K., Li, Q., and Mawatari, K.: JHEP **0907**, (2009); Frank, J., Rauch, M., and Zeppenfeld, D.: Phys.Rev. **D87**, 055020 (2013); Englert, C., Gonsalves-Netto, D., Mawatari, K., and Plehn, T.: JHEP **1301**, 148 (2013).

[62] Djouadi, A., Godbole, R., Mellado, B., Mohan, K.: Phys. Lett. **B723**, 307 (2013).

[63] Barger, V., et al.: Phys. Rev. **D49**, 79 (1994); Grzadkowski, B., Gunion, J., and He, X.: Phys. Rev. Lett. **77**, 5172 (1996); Gunion, J., and Pliszka, J.: Phys. Lett. **B444**, 136 (1998); Bhupal Dev, P., et al.: Phys. Rev. Lett. **100**, 051801 (2008).

[64] Freitas, A., and Schwaller, P.: Phys. Rev. **D87**, 055014 (2013).

[65] Barbieri, R., and Giudice, G.: Nucl. Phys. **B306**, 63 (1988).

[66] Papucci, M., Ruderman, J., and Weiler, A.: JHEP **1209**, 035 (2012).

[67] Carena, M., Heinemeyer, S., Wagner, C., and Weiglein, G.: Eur. J. Phys. **C26**, 601 (2003).

[68] Djouadi, A., and Quevillon, J.: arXiv:1304.1787 [hep-ph].

[69] Maiani, L., Polosa, A.D., and Riquer, V.: New J. Phys. **14**, 073029 (2012); Phys. Lett. **B718**, 465 (2012); Phys. Lett. **B724**, 274 (2013).

[70] Carena, M., et al.: Eur. Phys. J. **C73**, 2552 (2013).

[71] Arbey, A., et al.: Phys. Lett. **B708**, 162 (2012).

[72] Arbey, A., et al.: JHEP **1209**, 107 (2012).

[73] Among the vast literature on the subject, for the (mainly) early papers in the SUSY context, see, e.g.: Baer, H., Barger, V., and Mustafayev, A.: Phys. Rev. **D85**, 075010 (2012); Draper, P., Meade, P., Reece, M., and Shih, D.: Phys. Rev. **D85**, 095007 (2012); Buchmueller, O., et al.: Eur. Phys. J. **C72**, 2020 (2012); Akula, S., et al.: Phys. Rev. **D85**, 075001 (2012); Strege, C., et al.: JCAP **1203**, 030 (2012); Beskidt, C., et al.: JHEP **1205**, 094 (2012); Carena, M., et al.: JHEP **1207**, 175 (2012); Carena, M., Low, I., and Wagner, C.: JHEP **1208**, 060 (2012); Cahill-Rowley, M., et al.: Phys. Rev. **D86**, 075015 (2012); Lodone, P.: Int. J. Mod. Phys. **A27**, 1230010 (2012); Kadastik, M., et al.: JHEP **1205**, 061 (2012); Ellwanger, U.: JHEP **1203**, 044 (2012); King, S., Muhlleitner, M., and Nevzorov, R.: Nucl. Phys. **B860**, 207 (2012); Ghilencea, D.: Nucl. Phys. **B876**, 16 (2013); Cao, J., et al.: Phys. Lett. **B710**, 665 (2012); Aparicio, L., Cerdeno, D., and Ibanez, L.: JHEP **1204**, 126 (2012); Ellis, J., and Olive, K.: Eur. Phys. J. **C72**, 2005 (2012); Cao, J., et al.: JHEP **1203**, 086 (2012); Boudjema, F., and La Rochelle, G.D.: Phys. Rev. **D86**, 015018 (2012); Brummer, F., Kraml, S., and Kulkarni, S.: JHEP **1208**, 089 (2012); Hall, L., Pinner, D., and Ruderman, J.: JHEP **04**, 131 (2012); Arvanitaki, A., and Villadoro, G.: JHEP **02**,

144 (2012); Delgado, A., et al.: Eur. Phys. J. **C73**, 2370 (2013); Heinemeyer, S., Stal, O., and Weiglein, G.: Phys. Lett. **B710**, 201 (2012); Bechtle, P., et al.: Eur. Phys. J. **C73**, 2354 (2013); Drees, M.: Phys. Rev. **D86**, 115018 (2012).

[74] Djouadi, A., et al: (MSSM WG). hep-ph/9901246.

[75] Chamseddine, A., Arnowitt, R., and Nath, P.: Phys. Rev. Lett. **49**, 970 (1982); Barbieri, R., Ferrara, S., and Savoy, C.: Phys. Lett. **B119**, 343 (1982); Hall, L., Lykken, J., and Weinberg, S.: Phys. Rev. **D27**, 2359 (1983).

[76] For a review see Giudice, G.F., and Rattazzi, R.: Phys. Rep. **322**, 419 (1999).

[77] Randall, L., and Sundrum, R.: Nucl. Phys. **B557**, 79 (1999); Giudice, G., Luty, M., Murayama, H., and Rattazzi, R.: JHEP **9812**, 027 (1998).

[78] Arkani-Hamed, N., and Dimopoulos, S.: JHEP **06**, 073 (2005); Giudice, G., and Romanino, A.: Nucl. Phys. **B699**, 65 (2004); Wells, J.: Phys. Rev. **D71**, 015013 (2005).

[79] See Hall, L., and Nomura, Y.: JHEP **1003**, 076 (2010); Giudice, G., and Strumia, A.: Nucl. Phys. **B858**, 63 (2012).

[80] Djouadi, A., Kneur, J.L., and Moultaka, G.: Comput. Phys. Commun. **176**, 426 (2007); Muhlleitner, M., et al.: Comput. Phys. Commun. **168**, 46 (2005).

[81] Heinemeyer, S., Hollik, W., and Weiglein, G.: Comput. Phys. Com. **124**, 76 (2000).

[82] For a review, see Craig, N.: arXiv:1309.0528 [hep-ph].

[83] For recent studies, see: Benhenni, A., et al.: Phys. Rev. **D84**, 075015 (2011); Li, T., et al.: Phys. Lett. **B710**, 207 (2012).

[84] Djouadi, A., Ellwanger, U., and Teixeira, A.M.: Phys. Rev. Lett. **101**, 101802 (2008); JHEP **0904**, 031 (2009).

[85] Ellis, J., et al.: Phys. Rev. **D70**, 055005 (2004).

[86] e.g., AbdusSalam, S.S., et al.: Eur. Phys. J. **C71**, 1835 (2011).

[87] For a review: Hurth, T., and Mahmoudi, F.: Rev. Mod. Phys. **85**, 795 (2013).

[88] See, e.g., Kane, G., Lu, R., and Zheng, B.: Int. J. Mod. Phys. **A28**, 1330002 (2013).

[89] Bernal, N., Djouadi, A., Slavich, P.: JHEP **0707**, 016 (2007).

[90] Delgado, A., and Giudice, G.: Phys. Lett. **B627**, 155 (2005); Arganda, E., Diaz-Cruz, J., Szynkman, A.: Eur. Phys. J. **C73**, 2384 (2013); Phys.Lett. **B722**, 100 (2013).

[91] See, e.g., Carena, M., et al.: Nucl. Phys. **B577**, 88 (2000).

[92] Djouadi, A., et al.: Phys. Lett. **B435**, 101 (1998); Eur. Phys. J. **C1**, 149 (1998); Eur. Phys. J. **C1**, 163 (1998).

[93] Spira, M., et al.: Nucl. Phys. **B453**, 17 (1995).

[94] Harlander, R., and Kilgore, W.: Phys. Rev. **D68**, 013001 (2003); Harlander, R., Liebler, S., and Mantler, H.: Comp. Phys. Comm. **184**, 1605 (2013).

[95] Arbey, A., Battaglia, M., and Mahmoudi, F.: Phys.Rev. **D88**, 015007 (2013); Bechtle, P., et al.: arXiv:1305.1933; Craig, N., Galloway, J., and Thomas, S.: arXiv:1305.2424; Christensen, N., Han, T., Su, S.: Phys. Rev. **D85**, 115018 (2012).

[96] The ATLAS collaboration: arXiv:1211.6956.

[97] The CMS collaboration: CMS-PAS-HIG-12-050.

[98] The CMS collaboration: CMS-PAS-HIG-12-033.

[99] The ATLAS collaboration: ATLAS-CONF-2013-090.

[100] The CMS collaboration: arXiv:1205.5736.

[101] ATLAS Coll.: arXiv:1307.7292; CMS Coll.: arXiv:1307.7135.

[102] Dawson, S., et al.: arXiv:1310.8361 [hep-ex].

[103] See, e.g., Baglio, J., et al.: JHEP **1304**, 151 (2013); Dolan, M., Englert, C., and Spannowsky, M.: JHEP **1210**, 112 (2012); Yao, W.: arXiv:1308.6302 [hep-ph].

[104] Abramowicz, H.: arXiv:1307.5288; Bicer, M., et al.: arXiv:1308.6176; Baer, H., et al.: arXiv:1306.6352; Brau, J., et al.: arXiv:1210.0202. For earlier work, see: Arons, G., et al.: arXiv:0709.1893; Aguilar-Saavedra, J.: hep-ph/0106315; Accomando, E., et al.: Phys. Rep. **299**, 1 (1998); Peskin, M., and Murayama, H.: Ann. Rev. Nucl. Part. Sci. **46**, 533 (1996); Djouadi, A.: Int. J. Mod. Phys. **A10**, 1 (1995).

Abdelhak Djouadi
LPT – Bât. 210
CNRS & Université Paris-Sud
F-91405 Orsay Cedex, France
e-mail: Abdelhak.Djouadi@cern.ch

Printed in the United States
By Bookmasters